Identification and Quantitation of DNA Adducts Derived From DBPs

The mission of the Awwa Research Foundation (AwwaRF) is to advance the science of water to improve the quality of life. Funded primarily through annual subscription payments from over 1,000 utilities, consulting firms, and manufacturers in North America and abroad, AwwaRF sponsors research on all aspects of drinking water, including supply and resources, treatment, monitoring and analysis, distribution, management, and health effects.

From its headquarters in Denver, Colorado, the AwwaRF staff directs and supports the efforts of over 700 volunteers, who are the heart of the research program. These volunteers, serving on various boards and committees, use their expertise to select and monitor research studies to benefit the entire drinking water community.

Research findings are disseminated through a number of technology transfer activities, including research reports, conferences, videotape summaries, and periodicals.

Identification and Quantitation of DNA Adducts Derived From DBPs

Prepared by:
Elmarie Bodes, **Jun Nakamura**, **David K. La**, and **James A. Swenberg**
University of North Carolina
Chapel Hill, NC 27599

Sponsored by:
Microbial/Disinfection By-Products Research Council

Jointly funded by:
Awwa Research Foundation
6666 West Quincy Avenue, Denver, CO 80235-3098
and

U.S. Environmental Protection Agency
1200 Pennsylvania Avenue, N.W., Washington, DC 20460

Published by:

Awwa Research Foundation
Advancing the Science of Water®

IWA Publishing

DISCLAIMER

This study was funded by the Awwa Research Foundation (AwwaRF). AwwaRF assumes no responsibility for the content of the research study reported in this publication or for the opinions or statements of fact expressed in the report. The mention of trade names for commercial products does not represent or imply the approval or endorsement of AwwaRF. This report is presented solely for informational purposes.

Published by IWA Publishing, Alliance House, 12 Caxton Street, London SW1H 0QS, UK

Telephone: +44 (0) 20 7654 5500; Fax: +44 (0) 20 7654 5555; Email: publications@iwap.co.uk;
Web: **www.iwapublishing.com**

AwwaRF report number 90968F. Originally published by AwwaRF for its subscribers in 2003.
IWA Publishing version published 2004.

ISBN 1 84339 873 7

CONTENTS

LIST OF TABLES

LIST OF FIGURES

FOREWORD

The Awwa Research Foundation is a nonprofit corporation that is dedicated to the implementation of a research effort to help utilities respond to regulatory requirements and traditional high-priority concerns of the industry. The research agenda is developed through a process of consultation with subscribers and drinking water professionals. Under the umbrella of a Strategic Research Plan, the Research Advisory Council prioritizes the suggested projects based upon current and future needs, applicability, and past work; the recommendations are forwarded to the Board of Trustees for final selection. The foundation also sponsors research projects through the unsolicited proposal process; the Collaborative Research, research Applications, and Tailored Collaboration programs; and various joint research efforts with organizations such as the U.S. Environmental Protection Agency, the U.S. Bureau of Reclamation, and the Association of California Water Agencies.

This publication is a result of one of these sponsored studies, and it is hoped that its findings will be applied in communities throughout the world. The following reports serves not only as a means of communication the results of the water industry's centralized research program but also as a tool to enlist the further support of the nonmember utilities and individuals.

Projects are managed closely from their inception to the final report by the foundation's staff and large cadre of volunteers who willingly contribute their time and expertise. The foundation serves a planning and management function and awards contracts to other institutions such as water utilities, universities, and engineering firms. The funding for this research effort comes primarily from the Subscription Program, through which water utilities subscribe to the research program and make an annual payment proportionate to the volume of water they deliver and consultants and manufacturers subscribe based on their annual billings. The program offers a cost-effective and fair method for funding research in the public interest.

A broad spectrum of water supply issues is addressed by the foundation's research agenda: resources, treatment and operations, distribution and storage, water quality and analysis, toxicology, economics, and management. The ultimate purpose of the coordinated effort is to assist water suppliers to provide the highest possible quality of water economically and reliably.

The true benefits are realized when the results are implemented at the utility level. The foundation's trustees are pleased to offer this publication as a contribution toward that end.

Edmund G. Archuleta, P.E.
Chair, Board of Trustees
Awwa Research Foundation

James F. Manwaring, P.E.
Executive Director
Awwa Research Foundation

PREFACE

The Microbial/Disinfection By-Products Research Council was established in 1995 as a vehicle for the selection and funding of research to provide scientific information in the areas of health effects, exposure assessment, risk assessment, and prevention and control of contamination by microbes and disinfection by-products in drinking water. The council is composed of representatives designated by the U.S. Environmental Protection Agency (USEPA), the Awwa Research Foundation (AwwaRF) Board of Trustees, the Association of State Drinking Water Administrators, the National Resources Defense Council, the National Environmental Health Association, or their designees. Sources of funding for this research include the USEPA and AwwaRF, along with other interested parties. The council disburses these funds for research deemed to be of the highest urgency and importance in resolving critical research issues in drinking water.

ACKNOWLEDGMENTS

The authors of this report are indebted to the following individuals for their cooperation and participation in this project:

Dr. Douglas Wolf for supplying the Eker rat tissues and large quantities of MX. Dr. Anthony DeAngelo for supplying the F344 rat and B6C3F1 mice tissues after exposure to DCA via drinking water. Patricia Upton for reviewing the final report and Rebecca Freeman and Ezekiel Chen, two undergraduate students in the laboratory, for their help in looking up references and typing documents. In addition, we would like to thank the Project Advisory Committee (PAC)--including Rodger Baird, Dr. Gary Boorman, Dr. Jeffrey Ross and Dr. Michael Plewa--and the AwwaRF Project Manager, Dr. Jarmila Popovicova, for their advice and guidance throughout this project.

EXECUTIVE SUMMARY

Chlorination of drinking water during disinfection generates a number of halogenated by-products with potential health risks. The spectrum of by-products and their concentrations in water varies, depending on the humic acid content, total organic carbon, and inorganics in the water. Many studies have been conducted to assess the risks associated with chlorinated drinking water and the major disinfection by-products. Epidemiologic studies suggest increased cancer risks associated with the consumption of chlorinated drinking water. These studies have related colon, rectum, and urinary bladder cancers to water sources containing elevated levels of chlorination by-products. It is unclear which disinfection by-products pose the greatest health risks. Attention has generally centered on the most prevalent compounds. Four trihalomethanes and two of the chlorinated acetates have been evaluated in chronic rodent bioassays and shown to induce cancer. At present, bromodichloromethane (BDCM) and dichloroacetate (DCA) are the major focus of attention because of their relatively high concentrations in water and their carcinogenic potencies. Other by-products, such as halogenated hydroxyfuranones, have not been studied extensively, but may be important sources of risk. Although extensive investigations of these chemicals have been conducted, mechanisms underlying their carcinogenic effects remain poorly understood. The *major objectives* of this research were to: (1) Characterize and quantitate DNA adducts derived from disinfection by-products (DBPs), including BDCM and chlorinated hydroxyfuranones. (2) Study the effects of exposure to DBPs on endogenous DNA adducts. (3) Examine the formation and type of abasic sites in DNA that result from exposure to DBPs.

The use of *DNA adducts as biomarkers of exposure* is a promising new technology. DNA adducts have biological relevance, as they are involved in the initiation and progression of carcinogenesis. The occurrence of DNA adducts is a good indicator of genetic damage from exposure to mutagens and carcinogens. DNA adducts at target sites should represent the molecular dose and provide critical information about exposure. This may be particularly important when exposure data are unavailable or difficult to determine, as is often the case in epidemiologic studies. DNA adducts also can be useful as biomarkers of effect and provide early indications of biological damage. Carcinogenicity tests in animals are generally conducted at high doses that greatly exceed potential human exposures making the relevance of such studies toward human health unclear. Measurement of DNA adducts integrates individual and species differences in toxicokinetic factors that could result in nonlinearities that affect the quantitative relationship between exposure and molecular dose. As a result, DNA adducts provide a mechanistic basis for conducting high-to-low dose, route-to-route, and interspecies extrapolations. For these reasons, the use of DNA adducts is likely to produce more accurate estimates of risks than current methods that are based on mathematical extrapolation of data from animal studies. There are many DNA adducts associated with endogenous processes. Rather than identifying and measuring a large number of endogenous DNA adducts, we decided to limit analyses to several important adducts. The hydroxyl radical-derived DNA adduct, *8-hydroxydeoxyguanosine* (8-OHdG), as an indicator of damage from oxidative stress, and the etheno DNA adducts, 1,N^6-ethenodeoxyadenosine (1,N^6-εdA), 3,N^4-ethenodeoxycytosine (3,N^4-εdC), and N^2,3-ethenoguanine (N^2,3-εGua), were measured as indicators of damage from lipid peroxidation. 8-OHdG was quantitated by HPLC-ECD, εdA and εdC were analyzed by immunoaffinity/^{32}P-postlabeling, and εGua by immunoaffinity chromatography coupled with gas chromatography/high resolution mass spectrometry. We expected that high-dose gavage administration of drinking water disinfectants would enhance concentrations of endogenous DNA adducts.

Currently, 8-OHdG is one of the most popular markers for the evaluation of oxidative DNA damage. Commonly, the adduct 8-OHdG has been measured by non-selective enzymatic digestion of DNA to nucleosides followed by high performance liquid chromatography with electrochemical detection (HPLC-ECD). 8-OHdG has biologic significance since it induces G:C→T:A transversions during DNA replication, and it plays a role in senescence, carcinogenesis and various disease processes. Considerable progress is also being made with the development of a new immuno slot-blot method for detecting oxidized purines in DNA by using the Ogg1 enzyme in combination with the AP site assay. 8-OHdG is repaired by Ogg1 which is a bifunctional DNA glycosylase that cleaves the *N*-glycosylic bond in DNA leaving an AP site.

Apurinic/apyrimidinic (AP) sites are very common DNA lesions that result from enzymatic cleavage of the *N*-glycosylic bond between deoxyribose and purines or pyrimidines, or by spontaneous depurination. DNA glycosylases recognize a specific abnormal base and remove it from DNA leaving an apurinic/apyrimidinic (AP) site and an intact phosphodiester bond. The base loss may be spontaneous or can be enhanced by alkylating agents or physical treatments (i.e. irradiation) which cause base modifications, a weakening of the glycosyl bond and an increase in the release of bases. AP sites are constantly repaired by the base excision repair pathway. However, a fraction of AP sites can escape repair causeing cell toxicity or lethality and inhibition of DNA replication due to the formation of strand breaks. AP sites also induce mutations, chromosome aberrations and transcription errors. Our laboratory developed a highly sensitive and specific slot-blot assay, which enables the detection of 2.4 AP sites/10^7 bases. The slot-blot assay measures the amount of heat-labile DNA adducts as well as AP sites generated by DNA glycosylase activity.

Evidence suggests that *etheno DNA adducts* play a role not only in chemical induced tumorigenesis, but also in several human cancers in which persistent oxidative stress leads to malignancy by increasing mutation rates and genomic instability. Exocyclic etheno adducts (N^2,3-εGua; 1,N^2-εGua; 1,N^6-εdA; 3,N^4-εdC) are formed endogenously as a result of lipid peroxidation, as well as after exposure to carcinogens like vinyl chloride. Etheno adducts are suspected of playing significant roles in oxidative stress as endogenous mutagens. A sensitive and selective assay was developed in our laboratory for N^2,3-εGua, 1,N^2-εGua and 1,N^6-εdA quantitation using immunoaffinity chromatography coupled with mass spectrometry. The incorporation of the immunoaffinity enrichment procedure during quantitation of the etheno adducts lessens the potential for contamination or artifactual adduct formation during sample preparation. Mass spectrometry is an important tool for identification and characterization of DNA adducts, and it allows detection of attomole quantities of DNA adducts.

Many chemicals, including several disinfection by-products, are carcinogenic in experimental animals when administered by gavage. When animals are exposed by other routes, such as drinking water, many of these chemicals are not carcinogenic, or induce tumors at significantly lower incidences. There is considerable debate over the relevance of high-dose gavage studies to human health, since human exposure to chemicals differs markedly from the mouse and rat exposed by gavage. Potential human exposures to disinfection by-products would occur by drinking water at lower concentrations. Based on consumption of 2 liters of water, potential human exposure would range from 0.1 to 0.3 μg/kg, several orders of magnitude lower than exposures in animal studies. Gavage administration of chemicals result in bolus doses, whereas human exposure routes associated with drinking water involve delivery over an extended period of time, which permits more efficient detoxication and elimination. Gavage administration is likely to produce greater concentrations of test substances at target sites. This difference can affect saturable processes, such as bioactivation, detoxication, and DNA repair, resulting in important quantitative differences in toxicity and carcinogenicity. One area of difference between gavage and drinking water exposure to chemicals may be in their effects on endogenous DNA adducts. Bolus exposure to chemicals is likely to deplete cellular defenses, and enhance concentrations of endogenous DNA adducts derived from reactive oxygen species and lipid peroxidation. One would not expect a significant increase in endogenous DNA adducts to occur with drinking water exposure.

We selected four chemicals to determine their effect on DNA adduct and AP site formation: (a) *Bromodichloromethane (BDCM):* It has been shown that BDCM is metabolized to intermediates that covalently bind to lipids and proteins and this binding has been associated with renal and hepatic toxicity in rodents. Evidence suggests that BDCM-induced genotoxicity is derived from its conjugation with glutathione. It is widely known that glutathione conjugate formation is an important bioactivation mechanism for several groups of compounds with implications for organ-selective toxicity and carcinogenicity. For example, vicinal dihaloalkanes such as 2,3-dihaloethanes yield conjugates that can give rise to highly electrophilic episulphonium ions that are involved in the cytotoxicity and mutagenicity of 1,2-dihaloethanes. Our objective was to synthesize, purify and characterize the dihalomethyl-glutathione conjugate formed between BDCM and glutathione. After a complete characterization of the conjugate product by HPLC, MS and NMR, we proposed to analyze DNA from exposed animals to determine whether this adduct(s) is also formed *in vivo*. The identification of BDCM-induced DNA adducts would indicate whether glutathione is involved in its activation.

Initially it was thought that BDCM might cause direct-acting glutathione conjugates, like 1-bromo-2-chloroethane, that can form electrophilic episulphonium ions. However, this is not possible since BDCM contains only a single carbon atom. It is more likely that BDCM forms direct toxic glutathione *S*-conjugates like halogenated alkanes (e.g. dichloromethane). We were not able to detect any conjugates of BDCM and glutathione. Both the HPLC and mass spectrometric determinations could not demonstrate such compounds. This could result from a reaction product that is labile, which is quite different than that of haloethane-glutathione conjugates. The reactions between a BDCM-glutathione mixture and adenosine resulted in a product with UV and MS spectra that were suggestive of an adduct, with detected ions being different from those in control samples. However, the potential DNA adduct could not be further characterized because of the low yield in which it was produced. We were also unable to identify a possible chemical structure that was consistent with the MS data. Nevertheless, this observation was consistent with the hypothesis that a reactive, short-lived compound such as *S*-(1-dihalomethyl)glutathione conjugate could be responsible for DNA adduct formation.

We also examined whether high-dose exposure to BDCM affects DNA lesions associated with oxidative stress and lipid peroxidation. We exposed male F344 rats to BDCM (0, 0.3, 1, or 3 mmol/kg, n=5 per group) by corn oil gavage. Rats were sacrificed 6 hours following a single exposure. Hepatic DNA was analyzed for 8-OHdG by HPLC-ECD. We did not detect an increase in 8-OHdG following exposure to BDCM. In addition, there was no difference in AP site accumulation from liver DNA in any of the dose groups when compared to the control group.

(b) *Dichloroacetate (DCA)*: Among haloacetates, only DCA and trichloroacetate (TCA) have been evaluated in animal carcinogenesis bioassays. DCA induces hepatocellular adenomas and carcinomas in mice and rats, whereas TCA induces hepatocellular adenomas and carcinomas only in mice. The induction of hepatic tumors by TCA is closely associated with peroxisome proliferation. Species differences in this response may explain the greater sensitivity of mice compared to rats. DCA also has some activity as a peroxisome proliferator, but induces tumors at doses below those required to induce peroxisome proliferation. Previous research had shown that haloacetates induce lipid peroxidation and oxidative DNA damage when administered by gavage. Brominated haloacetates induced prolonged oxidative stress in the livers of B6C3F1 mice and acute exposures to the brominated derivatives showed that they were more potent inducers of oxidative damage to DNA (increase in 8-OHdG), and lipid peroxidation (measured by thiobarbituric acid-reactive substances; TBARS) than the chlorinated forms. The bromine-carbon bond is more labile than the chlorine-carbon bond which may explain why brominated haloacetates showed a greater increase in lipid peroxidation and 8-OHdG formation. Since it is already known that dibromoacetate induces oxidative stress we decided to focus our efforts on DCA in stead. In addition, liver tissue form mice exposed to DCA became available to our research group and thus our goal was to determine if AP site formation was induced after exposure to DCA in drinking water. We also investigated whether DCA induces the formation of lipid peroxidation-induced etheno DNA adducts. Male F344 rats and B6C3F1 mice were exposed to DCA via drinking water for 28 days at 0, 0.5 or 1.5 g/L for rats, and 0 or 3.5 g/L for mice. We analyzed hepatic DNA for 8-OHdG by HPLC-ECD and we examined liver DNA for the lipid peroxidation induced DNA adducts by ^{32}P-postlabeling. We did not observe an increase in these endogenous DNA lesions. The mean values in many instances are suggestive of some cause and effect relationships, but variance in the data sets perhaps prevents the ability to make salient observations. There are large variances in the 1,N^6-εdA and 3,N^4-εdC results which are inherent to the ^{32}P-postlabeling method and repeated measurements are unlikely to overcome this problem. Other approaches, such as mass spectrometry, which yield results with greater reproducibility are likely to reduce the variance and increase the likelihood of achieving statistical significance. We currently have a GC/MS method for the quantitation of N^2,3-εGua and 1,N^2-εGua, and a LC/MS method for the quantitation of 1,N^6-εdA. However, no additional liver tissue or DNA were available for determining adduct levels by mass spectrometry. In addition, we measured AP sites in hepatic DNA of the exposed mice. There was no significant difference between the control and exposed groups.

(c) *Mutagen X (MX):* MX is a potent, direct-acting mutagen that can account for as much as 60% of the mutagenic activity in chlorinated drinking water. The mechanism underlying the genotoxicity of MX in *Salmonella* is not fully understood, and little is known about the chemical reactions between MX and DNA. Because of its potent mutagenicity, there is considerable interest in identifying DNA adducts of MX. The objective of our research was to identify DNA adducts of MX using HPLC electrospray ionization tandem mass spectrometry (HPLC-ESI/MS/MS). Since MX is direct acting, we decided to incubate MX with calf thymus DNA as well as with individual deoxyribonucleosides. We expected that HPLC-ESI/MS/MS would enable sensitive detection of MX-derived DNA adducts, as well as provide important structural information. In addition, an extensive and unique set of tissue samples became available for this project through a collaborative effort between our laboratory and the US EPA. This study was designed to examine the enhancing effect of individual and a mixture of drinking water disinfection by-products in a rat model of hereditary renal cell carcinoma (the Eker rat).

Because this animal model is well defined in regard to both lesion development and genetic mutation, it allows for direct quantitation of carcinogenic response and production of statistically powerful data to understand the relative toxicity of chemicals and genetic alterations that may be chemical specific. Eker rats were exposed via drinking water to low and high doses of individual or a mixture of renal carcinogens and/or nephrotoxicants for 4 months. We were particularly interested in the Eker rats that were exposed via drinking water to MX, at a low concentration 0.007 g/L and a high concentration of 0.07 g/L. We focused our attention on liver samples for 8-oxoguanine, AP sites and ethenoguanine adducts. We quantitated 8-oxoguanine by HPLC-ECD following selective enzymatic digestion of DNA by the base excision repair enzyme formamidopyrimidine glycosylase (Fpg). No detectable amounts of 8-oxoguanine were observed in liver DNA from control and exposed groups analyzed. We used the sensitive slot-blot assay to quantitate aldehydic DNA lesions in liver DNA from F344 rats (single gavage; 0 or 15 mg/kg) and Eker rats (drinking water; 4 months; 0 or 0.07 g/L). We found no significant difference between the control and exposed rats in both studies. In addition, neither N^2,3-εGua nor 1,N^2-εGua adducts were found in the *in vitro* reaction of MX with calf thymus DNA or in liver DNA following exposure to MX in drinking water for 4 months. These combined results suggest that exposure to MX at doses used in these studies do not induce oxidative stress. In addition, if ethenoadducts were formed by MX, the amounts formed were small or chemically different from those evaluated.

In order to further examine this issue, we decided to synthesize the ethenoformyldeoxyguanosine (εfdGuo) adduct previously suggested to be formed by MX on a large scale to be used in our future experiments. Our aim was to determine if such an adduct is formed in calf thymus DNA incubated with MX and in liver samples of rats exposed to MX via drinking water. Reactions were performed at 50°C to improve the yield. HPLC analysis indicated that the yield of the product improved from 2.75% to 4.15% calculated by the area of the 15.7 minute peak for the adduct, but the spectral data indicated that the isolated adduct was impure. After several months of extensive research, we were not able to develop a method that provided large amounts of relatively pure adduct to be used in quantitative studies for the synthesis of analyte standards and isotopically labeled internal standards. Subsequently, we were unable to develop a quantitative LC/ESI/MS/MS method to investigate the presence of εfdGuo adducts in calf thymus DNA exposed *in vitro* to MX and DNA from male Eker rats exposed to MX.

(d) *Mucochloric Acid (MCA):* Although MCA is not as abundant in drinking water as trihalomethanes and haloacetic acids, it has several interesting features that merit its investigation. First, MCA and other halogenated furanones are potent mutagens. In addition, several of the DNA adducts induced by MCA are also formed endogenously from lipid peroxidation. DNA adduct formation by MCA *in vitro* has been reported following the reaction of MCA with nucleosides, resulting in etheno and ethenocarbaldehyde derivatives. We had developed an assay for the detection and quantitation of the etheno DNA adduct, N^2,3-εGua using immunoaffinity (IA) chromatography coupled with gas chromatography/high resolution mass spectrometry (GC/HRMS).

MCA was reacted with deoxyguanosine and a large amount of N^2,3-εGua was produced from the reaction. The formation of this adduct from this reaction has not been previously reported, most likely due to the greater sensitivity of the GC/HRMS method. We also administered MCA to F344 rats at 100mg/kg. MCA was dissolved in DMSO, mixed with corn oil, and administered by gavage. Animals were sacrificed at 4, 12, or 24 hours following exposure. We analyzed hepatic DNA for increases in N^2,3-εGua, an adduct that is normally present in DNA from lipid peroxidation, using the newly developed GC/HRMS method. N^2,3-εGua was measured at similar concentrations in control and treated rats. Although a large amount N^2,3-εGua was formed from the *in vitro* reaction, the same adduct was not induced following *in vivo* exposure. The reason for this difference is not known. Our results highlight the importance of performing *in vivo* studies on suspect carcinogens such as MCA in drinking water in conjunction with *in vitro* studies.

In addition, we analyzed the hepatic DNA from rats exposed to MCA and there was a statistically significant increase in AP site formation only at the 24 hour time point. Subsequently, an acute dose response study was conducted where male F344 rats were exposed to a single dose of MCA at 0, 10, 30, 100 or 300 mg/kg dissolved in distilled deionized water via gavage. The major focus of this experiment was to measure the amount of endogenous AP sites in genomic DNA to better understand the formation and repair of AP sites at basal levels after exposure to MCA. Unexpectedly, 24h after treatment, all the rats in the high dose group and one of the 100 mg/kg animals were found dead. In addition there was no statistically significant difference, in the amount of AP sites, between the control and MCA exposed animals. We speculate that the different gavage vehicles used between this study and the previous study may explain the sudden death after exposure. In the first study MCA was dissolved in DMSO and mixed with corn oil while in the second study MCA was dissolved in water alone. We believe that the water vehicle allowed for much greater absorption of MCA across the gastrointestinal mucosa with subsequent detrimental effects.

Final Conclusions: Direct DNA adduct formation was examined by exposing BDCM, MCA or MX to nucleosides or DNA. While N^2,3-εGua readily formed *in vitro* with MCA, none was detectable in MCA-treated rat liver. The yield of DNA adducts formed *in vitro* with glutathione and BDCM was low. Likewise, MX did not readily form ethenoformyl guanine adducts *in vitro* and no adducts were detected in rats chronically exposed to MX.

We hypothesized that increased numbers of endogenous DNA adducts would be present in DBP-treated animals due to depletion of normal cellular defenses. Endogenous DNA adducts were examined in tissues from rodents exposed to MX, MCA, BDCM and DCA and controls. No increases in endogenous adducts were found. Abasic sites are non-coding lesions in DNA that arise following oxidative stress and during DNA repair. We examined tissues from rodents treated with BDCM, MX, DCA and MCA. No increases in AP sites were found in rodents treated with BDCM, MX, or DCA.

Several disinfection by-products have been shown to be carcinogenic in experimental animals. Likewise, epidemiology studies have suggested increased risks of cancer associated with drinking of chlorinated drinking water. The long-term objective of this research was to better understand the health risks associated with DBPs. We demonstrated that several carcinogenic DBPs do not form direct DNA adducts and do not enhance endogenous adducts. Thus, they are not similar to well known genotoxic carcinogens. This may have important implications for risk assessment.

CHAPTER 1: INTRODUCTION AND BACKGROUND

E. Bodes, J. Nakamura, B. F. Pachkowski, D.K. La and James A. Swenberg

OVERVIEW

Chlorination of drinking water during disinfection generates a number of halogenated by-products with potential health risks (IARC, 1991). The spectrum of by-products and their concentrations in water vary and depend on the humic acid content, total organic carbon, and inorganics in the water. For example, high levels of bromine in the water lead to the formation of brominated by-products. Trihalomethanes and haloacetates are generally among the most prevalent classes of disinfection by-products identified in drinking water in the United States. Their concentrations typically range from 1 to 50 μg/L. Many studies have been conducted to assess the risks associated with chlorinated drinking water and the major disinfection by-products.

Epidemiologic studies suggest increased cancer risks associated with the consumption of chlorinated drinking water (IARC, 1991). These studies have related colon, rectum, and urinary bladder cancers to water sources containing elevated levels of chlorination by-products. Several other cancer sites, including the stomach, brain, pancreas, lung, and liver, have also been related to chlorination by-products, but these associations appear less conclusive.

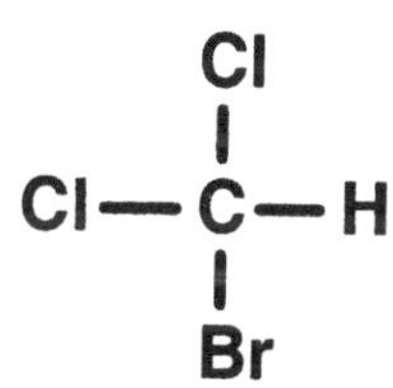

Bromodichloromethane (BDCM)

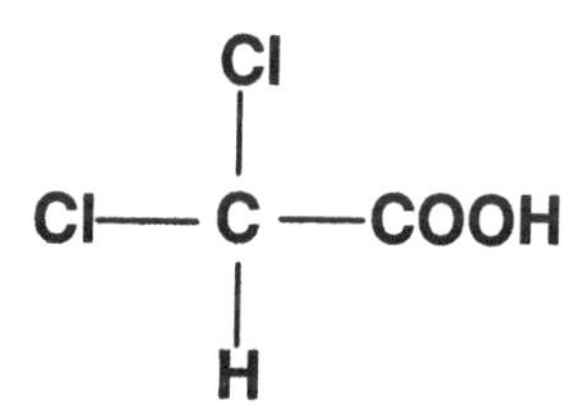

Dichloroacetate (DCA)

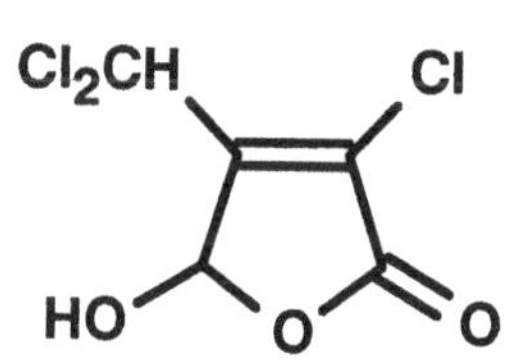

Mutagen X (MX)

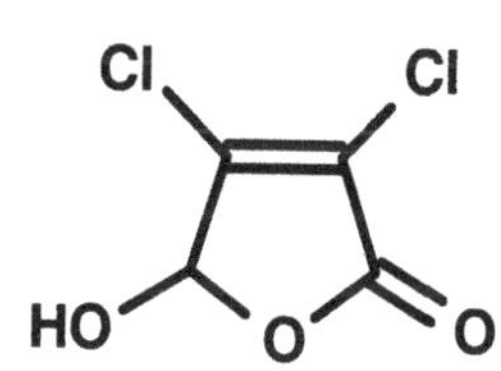

Mucochloric Acid (MCA)

Figure 1.1. Chemical structures of water disinfection by-products of relevance to AwwaRF research project.

It is unclear which disinfection by-products pose the greatest health risks. Attention has generally centered on the most prevalent compounds. Four trihalomethanes and two of the chlorinated acetates have been evaluated in chronic rodent bioassays and shown to induce cancer (IARC, 1991). At present, BDCM and DCA are the major focus of attention because of their relatively high concentrations in water and their carcinogenic potencies. Other by-products, such as halogenated hydroxyfuranones, have not been studied extensively, but may be important sources of risk.

Trihalomethanes

All four trihalomethanes [chloroform, chlorodibromomethane, BDCM (Figure 1.1), and bromoform] are carcinogenic in rodents in chronic gavage studies, but different mechanisms may be involved (IARC, 1991; Bull *et al.*, 1995). Chloroform is nongenotoxic, whereas the brominated trihalomethanes are genotoxic. The carcinogenicity of chloroform is hypothesized to occur secondary to its cytotoxic effects. Liver tumors in B6C3F1 mice and kidney tumors in Osborne-Mendel rats occur only at doses that induce cytotoxicity and regenerative cell proliferation. By contrast, BDCM carcinogenesis likely involves genotoxicity. BDCM induces adenocarcinomas of the large intestine and kidneys in F344 rats, renal adenocarcinomas in male mice, and hepatocellular adenomas and carcinomas in female mice (Dunnick *et al.*, 1987). Significant DNA binding occurs in these organs (Bull *et al.*, 1995). Glutathione *S*-transferase θ appears to mediate transformation of BDCM to mutagenic intermediates. It is interesting to note that approximately 30% of Caucasians are devoid of glutathione *S*-transferase θ (Schroder *et al.*, 1992; Pegram *et al.*, 1997) and would not be at risk via this mechanism. Therefore, this polymorphism may be an important factor in affecting individual susceptibility to disinfection by-products.

The carcinogenicity of trihalomethanes decreases significantly when administered by routes and methods other than corn oil gavage. For example, in drinking water studies of chloroform, only the highest dose (1800 ppm) produced a statistically significant increase in rat kidney tumors. Likewise, drinking water administration of BDCM induced only neoplastic nodules and adenofibrosis of the liver in rats. Changes in carcinogenicity with different exposure routes may relate to the dose rate. Bolus exposures are likely to produce greater doses at target sites and exceed the capacity of cellular defense mechanisms compared to more protracted exposures (La *et al.*, 1996). In turn, saturation of cellular defenses may lead to greater toxicity and may enhance the formation of endogenous DNA adducts from reactive oxygen species and lipid hydroperoxides.

Haloacetates

Among haloacetates, only dichloroacetate (DCA; Figure 1.1) and trichloroacetate (TCA) have been evaluated in animal carcinogenesis bioassays. DCA induces hepatocellular adenomas and carcinomas in mice and rats, whereas TCA induces hepatocellular adenomas and carcinomas only in mice (Bull *et al.*, 1995; IARC, 1995). The induction of hepatic tumors by TCA is closely associated with peroxisome proliferation. Species differences in this response may explain the greater sensitivity of mice compared to rats (DeAngelo *et al.*, 1989). DCA also has some activity as a peroxisome proliferator, but induces tumors at doses below those required to induce peroxisome proliferation (DeAngelo *et al.*, 1989; Daniel *et al.*, 1992).

Haloacetates induce lipid peroxidation and oxidative DNA damage when administered by gavage (Larson and Bull, 1992; Austin *et al.*, 1995, 1996). Brominated haloacetates produce greater amounts of the oxidized DNA lesions, such as 8-hydroxydeoxyguanosine, than DCA and TCA in mice exposed by gavage. Drinking water administration of brominated haloacetates also increases 8-hydroxydeoxyguanosine, whereas DCA and TCA fail to produce increases (Parrish *et al.*, 1996). Thus, it would appear that brominated haloacetates possess a greater ability to induce oxidative DNA damage than the chlorinated derivatives. Whether oxidative DNA lesions are involved in the carcinogenesis of haloacetates is not known. We are currently investigating whether DCA induces the formation of lipid peroxidation-induced etheno DNA adducts.

Halogenated hydroxyfuranones

While trihalomethanes and haloacetates are the predominant disinfection by-products, most of the mutagenicity associated with chlorinated drinking water derives from minor products, such as 3-chloro-4-(dichloromethyl)-5-hydroxy-2(*5H*)-furanone (MX; Figure 1.1) and related chlorinated hydroxyfuranones. MX is one of the most potent bacterial mutagens known, producing 5000 to 13,000 revertants per nmol in *Salmonella typhimurium* TA100 (DeMarini *et al.*, 1995). While its concentration in water is relatively low (1 to 60 ng/L), MX accounts for as much as 60% of the mutagenic activity of some chlorinated drinking water samples. The combined mutagenic activities of MX and its isomer E-MX (Figure 1.2), can account for more than 90% of the total mutagenicity of chlorinated drinking water. A recent 2-year carcinogenicity bioassay has shown MX to be carcinogenic in rats (Komulainen *et al.*, 1997).

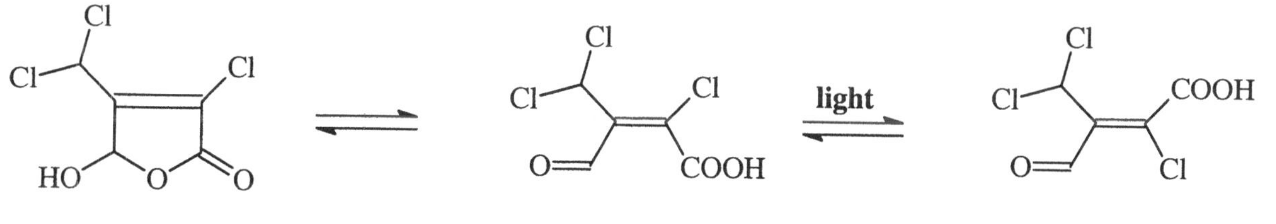

MX Open form E-MX

Figure 1.2. Structures of MX and E-MX.

MX reacts with DNA to form adducts, *in vitro* and *in vivo*, but specific adducts have not yet been characterized (Meier *et al.*, 1989; Schut *et al.*, 1991). However, DNA adducts of several other chlorinated hydroxyfuranones have been characterized. 3,4-Dichloro-5-hydroxy-2(5H)-furanone, or mucochloric acid (MCA; Figure 1.1), reacts with nucleosides under physiological conditions to form etheno and ethenocarbaldehyde derivatives of adenosine, cytidine, and guanosine (Kronberg *et al.*, 1992, 1993). Etheno DNA adducts are promutagenic lesions that are induced by a number of known carcinogens, including vinyl chloride and urethane, as well as by-products of lipid peroxidation (La and Swenberg, 1996). Whether or not MCA induces formation of these promutagenic DNA lesions *in vivo* needs to be investigated.

DNA ADDUCTS AS MARKERS OF EXPOSURE AND EFFECT

The use of DNA adducts as biomarkers of exposure is highly advantageous. DNA adducts have biological relevance, as they are involved in the initiation and progression of carcinogenesis (La and Swenberg, 1996). The occurrence of DNA adducts is a good indicator of genetic damage from exposure to mutagens and carcinogens. DNA adducts at target sites represent the molecular dose and can provide critical information about exposure. This may be particularly important when exposure data are unavailable or difficult to determine, as is often the case in epidemiologic studies. DNA adducts (Figure 1.3) also can be useful as biomarkers of effect and provide early indications of biological damage (Figure 1.3).

Carcinogenicity tests in animals are generally conducted at high doses that greatly exceed potential human exposures making the relevance of such studies toward human health unclear. Measurement of DNA adducts integrates individual and species differences in toxicokinetic and toxicodynamic factors, as well as any nonlinearities that affect the quantitative relationship between exposure and molecular dose. As a result, DNA adducts provide a mechanistic basis for conducting high-to-low dose, route-to-route, and interspecies extrapolations. For these reasons, the use of DNA adducts is likely to produce more accurate estimates of risks than current methods that are based on mathematical extrapolation of data from animal studies.

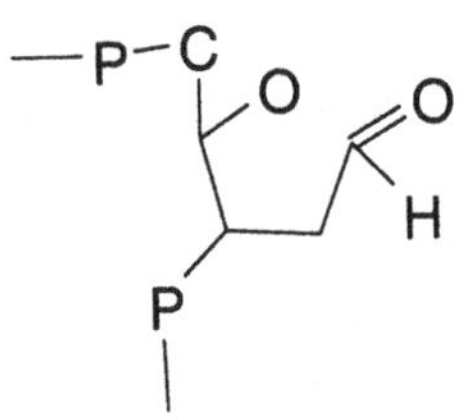

Apurinic/apyrimidinic (AP) Site

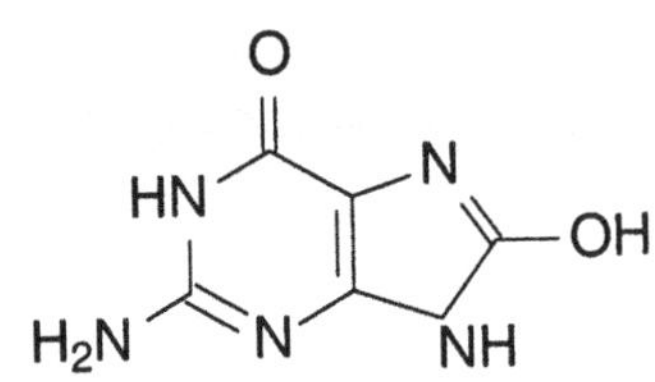

8-oxoguanine

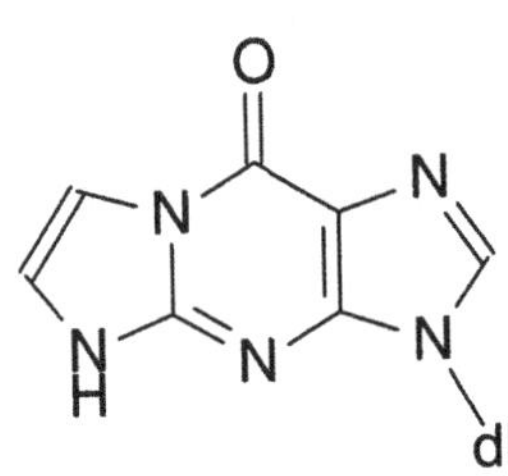

1,N^2-εdeoxyguanosine

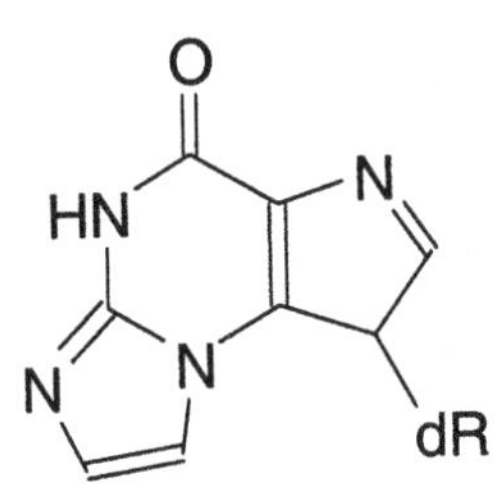

N^2,3-εdeoxyguanosine

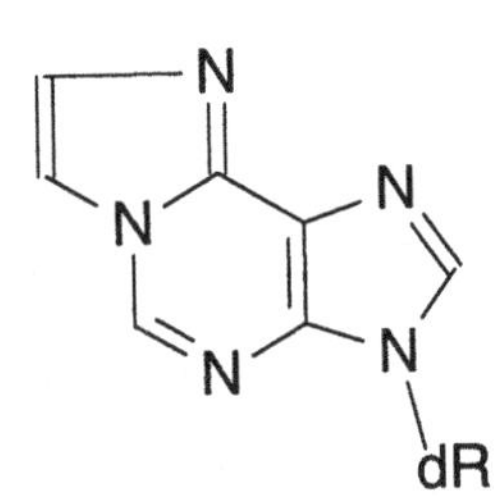

1,N^6-εdeoxyadenosine

Figure 1.3. Chemical structures of some DNA lesions derived from reactive oxygen species.

Biomarkers of oxidized DNA

Currently, 8-hydroxydeoxyguanosine (8-OHdG) is one of the most popular markers for the evaluation of oxidative DNA damage. Commonly, the adduct 8-OHdG has been measured by non-selective enzymatic digestion of DNA to nucleosides followed by high performance liquid chromatography with electrochemical detection (HPLC-ECD). 8-OHdG has biologic significance since it induces G:C→T:A transversions during DNA replication (Kasai, 1997; Floyed, 1990), and it plays a role in senescence, carcinogenesis and various disease processes. Despite popularity, there have been doubts regarding the accuracy of the 8-OHdG measurements because of the possibility of artifactual production during sample preparation.

Beckman *et al* (2000) proposed an easier and more robust method that promised to limit artificial oxidation. This assay is more selective with increased sensitivity, and the researchers concluded that this method is more efficient than the previous assay where non-selective nucleases induce incomplete DNA digestion. In this new protocol selective enzymatic digestion of DNA using formamidopyrimidine glycosylase (Fpg), an *Escherichia coli* base excision repair enzyme, releases the base 8-oxoguanine from full-length DNA. After digestion, 8-oxoguanine is separated from high molecular weight molecules, such as the DNA backbone, by ultrafiltration (10 kDa exclusion) and is then directly analyzed by HPLC-ECD. This ultra sensitive method was modified and validated in our laboratory using renal tissue samples from rats exposed to 0.4 g/L potassium bromate in drinking water (Detailed Materials and Methods) which is known to induce oxidative stress in rat kidneys (Umemura *et al.,* 1998).

Apurinic/Apyrimidinic (AP) sites (aldehydic DNA lesions)

AP sites are very common DNA lesions that result from enzymatic cleavage of the *N*-glycosyl bond between deoxyribose and purines or pyrimidines or by spontaneous depurination (Mol *et al.,* 2000). DNA glycosylases recognize a specific abnormal base and remove it from DNA leaving an apurinic/apyrimidinic (AP) site and an intact phosphodiester bond. The base loss may be spontaneous or can be enhanced by alkylating agents or physical treatments (i.e. irradiation) which cause base modifications, a weakening of the glycosyl bond and an increase in the release of bases (Talpaert Borle *et al.,* 1987). AP sites are constantly repaired by, for example, AP endonucleases during base excision repair. However, a fraction of AP sites that escapes repair causes cell toxicity or lethality and inhibits DNA replication due to the formation of strand breaks. AP sites also induce mutations, chromosome aberrations and transcription errors. AP sites in the DNA template hamper RNA polymerase and cause it to pause, which results in a decrease of RNA synthesis. AP sites may be by-passed by DNA polymerases resulting in preferential misincorporation of deoxyadenosine nucleotides. Class II APE are abundant in most cells and initiate repair by incising DNA immediately 5' to AP sites. Whereas APEs are DNA glycosylases associated with AP lyase activity which results in the incision of DNA immediately 3' to the AP site via β-elimination (Nakamura *et al.,* 2000; Demple *et al.,* 1997).

Our laboratory developed a highly sensitive and specific Slot-Blot assay, which enables the detection of 2.4 AP sites/10^7 bases (Nakamura *et al.,* 1998). This assay requires a small amount of genomic DNA (15 μg) and no radioactive materials. The slot-blot assay measures the amount of heat-labile DNA adducts as well as AP sites generated by DNA glycosylase activity.

Biomarkers of DNA adducts arising from lipid peroxidation

Evidence suggests that etheno-DNA adducts play a role not only in chemical induced tumorigenesis, but also in several human cancers in which persistent oxidative stress leads to malignancy by increasing mutation rates and genomic instability. Exocyclic etheno adducts

(N^2,3- and 1,N^2-ethenodeoxyguanosine; 1,N^6-ethenodeoxyadenosine; 3,N^4-ethenodeoxycytidine) are formed endogenously as a result of lipid peroxidation as well as after exposure to carcinogens like vinyl chloride (Morinello *et al.*, 2001). Etheno adducts are suspected of playing significant roles in oxidative stress as endogenous mutagens. A sensitive and selective assay was developed in our laboratory for N^2,3-ethenoguanine (N^2,3-εGua) quantitation using immunoaffinity (IA) chromatography coupled with gas chromatography/high resolution mass spectrometry (GC/HRMS; Ham *et al.*, 1999). This method was adapted by Morinello *et al.* (2001) to be used for the simultaneous quantitation of N^2,3-εGua and 1,N^2-εGua (1,N^2-ethenoguanine). Similarly, a fast and highly specific analytical method was developed in our laboratory to quantify 1,N^6-ethenodeoxyadenosine (1,N^6-εdA) by using IA chromatography combined with liquid chromatography/electrospray ionization with tandem mass spectrometry (LC/ESI-MS/MS).

The incorporation of the immunoaffinity enrichment procedure during quantitation of the etheno adducts lessens the potential for contamination or artifactual adduct formation during sample preparation. Mass spectrometry (MS) is an important tool for identification and characterization of DNA adducts, and it allows detection of attomole quantities of DNA adducts. Most quantitative MS methods monitor either a single or some prominent ions of the analyte, in addition to the stable isotope-labeled analog of the analyte which is used as internal standard (isotope dilution mass spectrometry). When electron capture negative chemical ionization (ECNCI) is used as the detection method there is a further increase in sensitivity and specificity for electrophoric compounds. Electrophore labeling (i.e. derivatization) of DNA adducts is necessary for detection by ECNCI.

ORIGINAL RESEARCH PLAN

The original research proposed by Dr. David K. La was divided into three phases. Phase 1 focused on identification of specific DNA adducts derived from disinfection by-products. Based on published studies, Dr. La et al expected many that disinfection by-products, including chloroform and haloacetates, would not bind covalently to DNA. The original research team planned to focus on BDCM and chlorinated hydroxyfuranones because DNA binding by these chemicals had been reported. Since MX accounts for most of the mutagenicity in drinking water, identification of MX-derived DNA adducts was a high priority. In Phase 2, the team investigated the affects of disinfection by-products on the formation of indirect DNA adducts. Dr. La et al expected that high-dose bolus exposure to chemicals would increase endogenous DNA adducts secondary to the saturation of cellular defenses. The extent to which different trihalomethanes and haloacetates enhanced concentrations of endogenous DNA adducts were compared, including 8-hydroxydeoxyguanosine and etheno adducts associated with lipid peroxidation. The original research team also planned to examine the affects of different exposure routes on endogenous DNA adducts. Based on the findings from Phases 1 and 2, the identification of appropriate biomarkers was planned for study in Phase 3.

Phase 1. Characterization and quantitation of DNA adducts derived from disinfection by-products.

Specific Aim 1. Identify DNA adducts derived from glutathione conjugates of BDCM and to quantitate their concentrations in tissues of exposed rats and mice.

Specific Aim 2. Identify DNA adducts derived from MX.
Specific Aim 3. Quantitate concentrations of etheno DNA adducts in rats exposed to MCA.

Phase 2. Study the effects of disinfection by-products on endogenous DNA adducts.

Specific Aim 4. Investigate whether disinfection by-products enhance concentrations of endogenous DNA adducts.
Specific Aim 5. Compare the affects of different exposure routes on endogenous DNA adducts.

Phase 3. Measurement of DNA adducts and abasic sites as biomarkers of exposure and effects of disinfection by-products.

Specific Aim 6. Quantitate DNA adducts derived from BDCM, MX and MCA in rat lymphocyte DNA.
Specific Aim 7. Quantitate endogenous DNA adducts (8-hydroxydeoxyguanosine, etheno-deoxyadenosine) excreted in urine.
Specific Aim 8. Investigate the affects of disinfection by-products on the formation of abasic sites.

REVISED RESEARCH PLAN

Dr. David La was the principal investigator for this project from 1998 – 2000. Drs. Danielle Lemaire, Jun Nakamura, Hasan Koc and Ramaiah Sangaiah formed the research group working closely with him during that time. When David La left the University of North Carolina, Dr. James Swenberg became the principal investigator for the remainder of the funding period. Drs. Elmarie Bodes, Jun Nakamura, Hasan Koc, Karupiah Jayaraj, and Kevin McDorman were the researchers/postdoctoral trainees, and Brian Pachkowski and Eric Morinello the graduate students that worked together to complete the remainder of the project. In addition to conducting her own dissertation research, Elmarie Bodes was also responsible for compiling data and to write up progress reports and the final report. The current research group evaluated the research progress during the first two years at a combined meeting with the Project Advisory Committee (PAC) and Project Manager on November 14, 2000. Since the initial results obtained from the original proposal were not moving forward as expected the new research team was advised by the PAC to change the direction of the research to focus on additional specific aims.

The scope of the research project was changed to focus on an extensive and unique set of tissue samples that became available for this project towards the end of the current funding period through a collaborative effort between our laboratory and Dr. Douglas Wolf with the National Health and Environmental Effects Research Laboratory at the US EPA. In this study, chemicals were selected on the basis of demonstrated nephrotoxicity and/or renal carcinogenesis in rodents. The rat model of hereditary renal cell carcinoma (the Eker rat) was used. This animal model is well defined in regard to both lesion development and genetic mutation, and it allows for direct quantitation of a carcinogenic response and the production of statistically powerful data to understand the relative toxicity of chemicals and genetic alterations that may be chemical specific. Male and female Eker rats were exposed via drinking water to MX for 20 or 45 weeks at low concentrations of 0.007 g/L and high concentrations of 0.07 g/L. The team focused its efforts on the analysis of frozen liver samples. Similarly, an additional 21 day exposure was conducted at the US EPA where Eker rats were exposed to BDCM, MX and $KBrO_3$ via drinking water at 0.7 g/L, 0.007 g/L and 0.4 g/L, respectively. This study focused on kidney samples. The team believed that these tissue samples from the chronically exposed rats in the EPA study would be valuable for gaining new information on chronic exposure to water disinfection by-products and that it would enhance the scope of the research project. In addition, exciting results were observed in a preliminary study where F344 rats were exposed to a single gavage dose of MCA at 100 mg/kg. In this study, there was an increase in AP sites in hepatic DNA examined at 24 h after dosage. An additional study was conducted to determine if a dose response existed for AP site accumulation after MCA treatment. This study was also not proposed in the original research proposal.

Our objectives for the Revised Research Plan were to identify several DNA adducts and lesions as biomarkers that relate to exposure and effect of MX, MCA and BDCM. The proposed research was divided into three main specific aims. In Specific Aim 9, we proposed to isolate and quantitate a MX specific ethenoformyl DNA adduct. In addition, we proposed to investigate other DNA adducts and lesions (i.e. 8-OHdG, AP sites and etheno adducts) in hepatic DNA from Eker rats after chronic exposure to MX via drinking water for 4 months. We expected that the HPLC-ESI/MS/MS method would be a very sensitive and specific tool to enable detection of the ethenoformyldeoxyguanosine DNA adduct derived from MX. We also believed that the IA-GC/HRMS method would be a powerful tool for the analysis of MX induced N^2,3-εGua and 1,N^2-εGua adducts due to extreme sensitivity and specificity. We speculated that high-dose chronic drinking water exposure would enhance concentrations of endogenous DNA adducts. For Specific Aim 10, we investigated several specific affects of MCA on F344 liver DNA in an acute dose response study. Specific Aim 11 focused on the identification of specific DNA adducts and lesions (i.e. 8-OHdG and AP sites) in Eker rat kidney DNA after exposure to BDCM via drinking water for 21 days. We expected that differences in toxicity would be reflected in differences in the extent of endogenous DNA adduct formation after chronic exposure. These additional objectives clearly added to and enhanced the objectives identified in the original research proposal. We expected that the IA-GC/HRMS method would be a powerful tool for the analysis of MCA induced N^2,3-εGua and 1,N^2-εGua adduct derived from lipid peroxidation. In addition we believed that the highly sensitive AP site assay would be a valuable tool to detect a commonly occurring DNA lesion (i.e. AP sites) and that dose-response study would enable us to identify doses at which thresholds occur in AP site accumulation.

CHAPTER 2: DETAILED MATERIALS AND METHODS

E. Bodes, J. Nakamura, B. F. Pachkowski, . E. J. Morinello, K. S. McDorman, D. Lemaire, H. Koc, P. Upton, D.K. La and James A. Swenberg

ANIMAL CARE

Animal care was in accordance with institutional guidelines of the University of North Carolina at Chapel Hill.

DNA ISOLATION

Tissues and organs were quickly harvested and frozen at -80°C until DNA extraction. Frozen tissues were thawed (0.3 – 0.5 g) and homogenized in 10 mL 1X PBS and 20 mM TEMPO mixture with a Tehran homogenizer (Wheaton Instruments, Millville, N.J.). After centrifugation at 2000 x g for 10 minutes, the nuclear pellets were incubated in lysis buffer (Applied Biosystems, Foster City, Calif.) overnight at 4°C with 125 μL proteinase K (Applied Biosystems, Foster City, Calif.). DNA was extracted twice with a mixture of phenol/chloroform/water and 20 mM TEMPO and once with a Sevag and 20 mM TEMPO mixture, followed by cold ethanol precipitation. The extracted DNA was incubated in 1.5 mL 1X PBS (pH 7.4) and 20 mM TEMPO mixture with 6 μL RNase A and 6 μL RNase T1. After DNA precipitation with cold ethanol, the DNA pellet was resuspended in deionized distilled water (ddH_2O) with 1 mM TEMPO. A 10 μL aliquot was used for measuring the UV absorbance on a spectrophotometer at the following wavelengths: 230, 260 and 280 nm. The remaining DNA solution was stored at -80°C until needed for further analysis.

QUANTITATION OF 8-HYDROXYDEOXYGUANOSINE BY HPLC-ECD

Following DNA isolation from tissues, DNA was enzymatically hydrolyzed using deoxyribonuclease, spleen phosphodiesterase, snake venom phosphodiesterase, and alkaline phosphatase (Richter *et al.,* 1988). Artifactual generation of 8-hydroxydeoxyguanosine (8-OHdG) was minimized during sample work-up the addition of radical scavengers (e.g. BHT or TEMPO). The digest was separated by reversed phase HPLC. 8-OHdG was quantitated by using an electrochemical array detector (ESA, Chelmsford, Mass.). Our electrochemical array system, equipped with multiple electrodes, significantly improved the sensitivity and specificity of detection. In a serial array of electrodes at increasing potential, compounds oxidized and measured at one electrode were effectively eliminated from detection by subsequent electrodes in the array. By this process, compounds which co-elute were clearly differentiated. Moreover, the electrochemical oxidation pattern at different potentials provided a "fingerprint" that facilitated qualitative characterization. Potentials were set at regular intervals between 300 and 900 mV. Quantitation was based on a standard curve generated from known quantities of 8-OHdG.

QUANTITATION OF 8-OXOGUANINE BY HPLC-ECD

We used a more direct approach for the identification of the oxidized DNA base, 8-oxoguanine. The original approach had measured the nucleoside of the adduct following non-selective enzymatic digestion of DNA. The new approach was based on a protocol published by Beckman *et al* (2000). This protocol involved the direct analysis of the oxidized base following selective enzymatic digestion of DNA using the *E. coli* base excision repair enzyme formamidopyrimidine glycosylase (Fpg). We modified this protocol and validated the assay in our laboratory.

As mentioned earlier artifactual generation of 8-oxoguanine was minimized during DNA isolation by the addition of radical scavengers (e.g. TEMPO; ICN Biomedicals Inc., Aurora, Ohio). Fpg enzyme stock (Bruce Ames, Berkeley, Calif.) was desalted on Bio-Rad Econo-Pac 10DG columns (Bio-Rad Laboratories Inc., Hercules, Calif.) and the concentration was quantified using the Bio-Rad protein assay. A 100 μg sample of isolated DNA was resuspended in 20 μL of 10X Fpg buffer, 6 μg of Fpg protein (0.35 μg/mL) was added and the volume of the hydrolysate was adjusted to a final volume of 200 μL with ddH_2O. The DNA-enzyme mixture was incubated for 1 hour at 37°C. The sample mixture was applied to a microcon-10 (10,000 Da; Millipore, Bedford, Mass.) vial and was centrifuged at 4°C and at 14,000 rpm for 45 minutes. The filtrate was loaded into HPLC injection vials and stored at -80°C until analyzed for 8-oxoguanine by HPLC-ECD (ESA Inc., Chelmsford, Mass.). Two hundred (200 μL) microliter of 0.1 N HCl was added to the Microcon filter containing the retentate (which contained the DNA backbone) and was allowed to incubate for 1 hour at 70°C. The hydrolysate was analyzed by HPLC-UV for quantitation of guanine. Data from 8-oxoguanine and guanine analysis was used to generate the number of 8-oxoguanine lesions per 10^6 guanine.

The digested filtrate was separated by reverse-phase HPLC with electrochemical detection and 8-oxoguanine was quantitated using an electrochemical array detector. A sample volume of 50 μL was injected onto the HPLC column. A 5 μm, 4 x 250 mm reversed phase C18 Beckman Ultrasphere ODS column (Beckman Coulter Inc., Fullerton, Calif.) was used. The column was eluted isocratically with 50 mM potassium phosphate monobasic buffer, pH 5.5 with 2.0% HPLC-grade methanol at a flow rate of 1 mL/min. The column was maintained at room temperature. The HPLC *ESA* liquid chromatographic system consisted of a coulometric electrode array detector (model 3600), model 580 pump, model 540 autosampler and CoulArray® for Windows (ESA Inc., Chlemsford, Mass.) data analysis program. The ECD consisted of a serial array of electrodes at increasing potential, which implied that compounds reacted and measured at one electrode were effectively eliminated from detection by subsequent electrodes in the array. Thus, compounds that co-elute were clearly differentiated since the electrochemical oxidation pattern provided a "fingerprint" that facilitates qualitative characterization. Electrochemical oxidation was monitored at 100, 110, 250, 350, 400, 450, 500 and 550 mV. The 8-oxoguanine peak was detected around 8 minutes and the signal was monitored at 250 mV. The concentration of 8-oxoguanine was normalized to the amount of DNA analyzed, as determined by UV absorbance.

The acid hydrolyzed samples were analyzed via HPLC-UV and guanine levels were quantitated. The HPLC-UV system consisted of a 510 Waters pump (Waters Corp., Milford, Mass.), an Applied Biosystems 757 UV absorbance detector, Waters 712 WISP autosampler and a Millennium^{32} Version 3.2 (Waters Corporation, Milford, Mass.) data analysis program. A Partisil (Whatman Inc., Clifton, N.J.) SCX (5μ, 4.6 x 250 mm) column was used for analysis and the column was eluted with an isocratic mobile phase (100 mM ammonium formate, 10% methanol water, pH 2.8 (stored at 4°C) at a flow rate of 1.8 mL/min. The following volumes of a guanine (42.1 μM guanine in 0.1 N HCl) stock solution were injected for the standard curve: 20, 30, 50, 100 and 150 μL. The UV detector was set at a wavelength of 254 nm. The average run time per sample was 5 minutes and the average observed retention time for guanine under these conditions was about 2.7 minutes.

In this new method we used the *Escherichia coli* repair enzyme formamidopyrimidine (Fapy) glycosylase to release the base 8-oxoguanine from full-length DNA. Since guanine is not released during the specific enzymatic hydrolysis by the *E. coli* Fapy repair protein, no additional adduct should have been generated. Enzymatic hydrolysis was followed by separating 8-oxoguanine from high molecular weight molecules by ultrafiltration and analyzing the filtrate for the base adduct by reverse-phase HPLC with electrochemical detection. The current limit of detection for 8-oxoguanine was five 8-oxo-dG per 10^7 guanine in a 100 μg DNA sample. In all calculations it was taken into consideration that the injection volume contains only 25 μg of DNA. Since the electrochemical detector can change slightly in a day-to-day response, calibration standards of 8-oxoguanine were required on each day of analysis. Although the UV detector displayed little day-to-day variation, calibration standards of guanine were also measured each day of analysis. The HPLC-ECD system was optimized for maximum sensitivity by reducing the background noise by incorporating a complete cleanup of the injection system prior to each run and running water blanks between samples to make sure there was no carry over between samples. Also, 8-oxoguanine standard (500 fmol) was injected after every six samples for quantitation. After collecting data from the standards, positive control and negative control (blank DNA) samples were injected prior to running the samples of interest. Afterwards we determined whether 8-oxoguanine was increased or not by comparing 8-oxoguanine data between control and treated groups that were analyzed on the same day with the same method. The positive controls used in these studies were obtained from kidney samples collected from previously exposed F344 rats (courtesy Dr. DeAngelo, US EPA). For further validation we used freshly isolated calf thymus DNA exposed to oxidative conditions to spike 8-oxoguanine standards, positive and negative control samples.

SLOT-BLOT ASSAY FOR APURINIC/APYRIMIDINIC (AP) SITES AND AP SITE CLEAVAGE ASSAY

This assay is based on the reaction between the aldehyde group in an AP site with a probe bearing a biotin residue (aldehyde reactive probe). The tagged biotin is quantitated by a slot-blot assay.

The AP site assay

This assay is performed by a procedure reported by Nakamura *et al.,* 2000. Briefly, 15 μg DNA in 150 μL PBS is incubated with 1 mM ARP (Dojindo Molecular Laboratories, Gaithersburg, Md.) at 37°C for 10 min. After precipitation using cold ethanol, DNA is resuspended in Tris-EDTA (TE)-buffer. The DNA concentration is measured by a UV spectrophotometer and the DNA solution is then prepared at 1.5 μg/100 μL in TE buffer. Heat-denatured DNA is immobilized on a nitrocellulose (NC) membrane (BA-85, Schleicher & Schuell, Piscataway, N.J.). The NC membrane is soaked with 5X sodium chloride/sodium citrate solution (SSC) and then baked in a vacuum oven at 80 °C for 30 min. The membrane is pre-incubated with 10 mL Tris-EDTA-NaCl buffer containing BSA for 15 min and then incubated in the same solution containing streptavidin-conjugated horseradish peroxidase (BioGen, Research Triangle Park, N.C.) at room temperature for 45 min. After rinsing the NC membrane, the enzymatic activity on the membrane is visualized by enhanced chemiluminescence reagents (Amersham Pharmacia Biotech, Piscataway, N.J.). The NC membrane is then exposed to X-ray film and the developed film is analyzed either by using an Ultrascan XL scanning densitometer or the new Kodak Image Analysis System.

AP site cleavage assay

The AP site cleavage assay is performed as described (Nakamura *et al.*, 2000) with a modification. *Determination of total AP sites*: The number of total AP sites are measured by the ASB assay as described above. *Quantitation of Exo III-excisable AP sites* (Figure 2.1): Five μg DNA is reacted with ARP followed by purification using a Microcon 50. The DNA is incubated with 145 IU *E. coli* exonuclease III (Exo III) (New England BioLabs, Beverly, Mass.) in 30 μL 50 mM HEPES/KOH buffer (pH 7.5) containing 50 mM NaCl, 5 mM $MgCl_2$ at 37°C for 1 min and is immediately analyzed by the ASB assay. The number of Exo III-excisable AP sites are calculated from the number of AP sites determined by the regular AP site assay minus the number of AP sites left after the treatment with Exo III. *Determination of T7-Exo-excisable AP sites* (Figure 2.1): Five μg DNA is reacted with ARP followed by purification as described above. The DNA is incubated with 200 U *T7* gene 6 exonuclease (T7 Exo) (USB Corp., Cleveland, Ohio) in 30 μL 50 mM HEPES/KOH buffer (pH 7.5) containing 50 mM NaCl, 5 mM $MgCl_2$ at 37°C for 1 min and immediately analyzed by the ASB assay. The number of T7-Exo-excisable AP sites are calculated from the number of AP sites determined by the regular AP site assay minus the number of AP sites left after the treatment with T7 Exo. *Calculation of intact and residual AP sites*: The numbers of residual AP sites are determined by the regular AP site assay minus the number of AP sites left after the treatment with Exo III and T7 Exo nucleases.

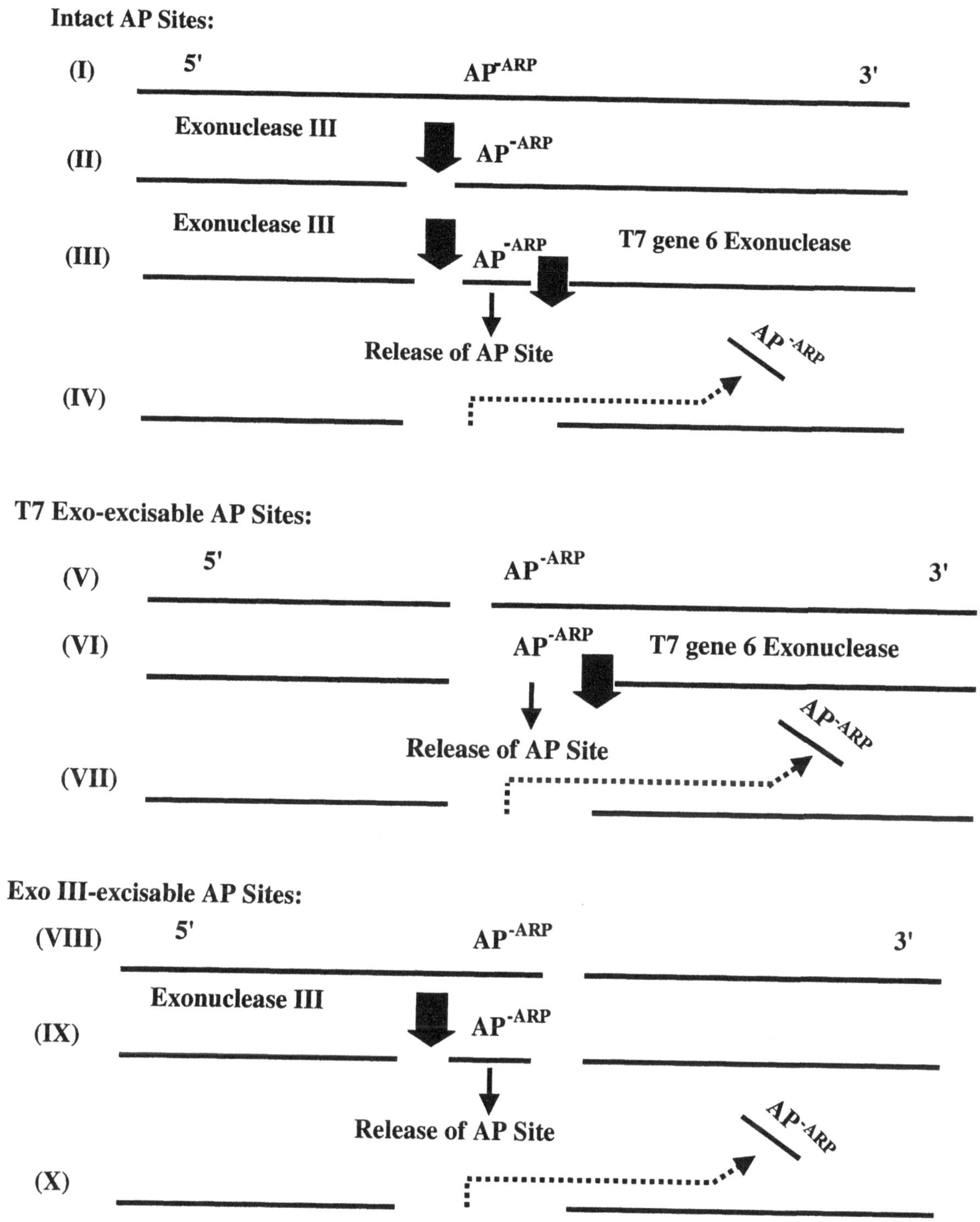

Figure 2.1. Quantitation of intact (steps I-III), T7 Exo-excisable (steps V-VII) and Exo III-excisable (steps VIII-X) AP sites in the AP site cleavage assay by using exonuclease III (Exo III) and T7 exonuclease (T7 Exo), respectively. Prepared by Jun Nakamura.

QUANTITATION OF ETHENOGUANINE (εGUA) ADDUCTS BY IA-GC/HRMS

A sensitive and selective assay was developed in our laboratory (Ham *et al.*, 1999) for N^2,3-ethenoguanine detection using immunoaffinity (IA) chromatography coupled with gas chromatography high resolution mass spectrometry (GC/HRMS). This method was adapted by Morinello *et al* (2001) to be used for the simultaneous quantitation of N^2,3-ethenoguanine and 1,N^2-ethenoguanine.

To each aqueous DNA (3-300 μg) or deoxyguanosine sample (10 μL of reaction solution) are added 66 fmol $[^{13}C_4,^{15}N_2]N^2$,3-εGua and 180 fmol $[^{13}C_3]1,N^2$-εGua internal standard. Water and 1 *N* HCl are added to yield a 1000 μL 0.1 *N* HCl solution. Each sample is incubated at 70 °C for 60 min, cooled on ice, and transferred to the reservoir of a Centricon-10 centrifugal filtration device. The samples are centrifuged at 5000 x *g* for 60 min to remove the DNA backbone. A 100 μL aliquot of each filtrate is removed and diluted with 150 μL water for guanine quantitation by HPLC. Each retentate is rinsed with 2 mL 500 mM phosphate buffer (pH 7.2) during a second 90 min centrifugation. The neutralized samples are diluted with 2 mL water and transferred to immunoaffinity columns constructed using polyclonal antibodies raised against each adduct (Figure 2.2). The columns are washed by sequential addition of 5 mL phosphate-buffered saline (pH 7.4), 0.02% sodium azide, 5 mL water, and 10 mL 5% methanol/water. The samples are eluted with 5 mL 100% methanol and are evaporated overnight under vacuum. The purified adducts are derivatized with pentafluorobenzyl bromide (PFBBr) to make them volatile for gas chromatography (GC). To each dried sample, 25 mg solid potassium carbonate, 500 μL acetone, and 35 μL 10 % PFBBr/acetone are added. The samples are incubated for 70 min at 50 °C followed by another evaporation under vacuum. Each sample is transferred in two 100 μL portions dichloromethane to disposable silica gel columns. The samples are washed with 4 mL hexane and 6 mL 5% ethyl acetate/hexane before elution in 3 mL ethyl acetate. The samples are evaporated and redissolved in 15 μL toluene for analysis. Adduct quantitation is performed using a HP5890 gas chromatograph fitted with a 30 m J&W DB-5MS capillary column and coupled to a Varian VG250SEQ II high-resolution mass spectrometer (HRMS). The HRMS is operated in the negative ion-chemical ionization mode, with a resolving power at 7-10 x 10^3. A 2 μL volume of each sample is injected. Selected ion monitoring at m/z = 334.0351 and 337.0453 are used to monitor the $[M-201]^-$ fragments of 3,5-PFB_2-1,N^2-εGua and 3,5-PFB_2-$[^{13}C_3]1,N^2$-εGua, respectively. The $[M-181]^-$ fragments of 3,5-PFB_2-N^2,3-εGua and 3,5-PFB_2-$[^{13}C_4,^{15}N_2]N^2$,3-εGua are monitored at m/z = 354.04 and 360.05, respectively. Each adduct is quantitated by comparison of its peak area with that of the corresponding internal standard. Adduct concentrations will be normalized to the amount of guanine measured in each sample by HPLC.

Figure 2.2. Immunoaffinity columns constructed using polyclonal antibodies raised against each adduct. After preparation columns are stored at 4 °C. Photo by Eric Morinello.

QUANTITATION OF THE ETHENOFORMYLDEOXYGUANOSINE (εFDGUO) ADDUCT BY LC/MS/MS

2'-Deoxyguanosine (2'-dGuo) was obtained from Sigma Aldrich Chemical Co. (St. Louis, Mo.) and MX was obtained from Dr. Wolf (US EPA, Research Triangle Park, N.C.). The purity of MX (7 mg dissolved in 0.75 mL solvent) and 2'-dGuo (3.5 mg dissolved in 0.75 mL solvent) were checked by ^{1}H-NMR (nuclear magnetic resonance) in DMSO-d_6. The purity of 2'-dGuo and MX was at least 98% as estimated by ^{1}H-NMR analysis.

Chromatographic methods

Analytical and preparative HPLC were performed on a liquid chromatographic system consisting of two model 510 Waters HPLC pumps (Waters Corp. Milford, Mass.), a 1040A-HPLC diode-array detector (UV; Hewlett-Packard, Palo Alto, Calif.), a Waters automated gradient control system and Hewlett Packard Chemstation for LC™ data analysis program. The reaction mixtures were chromatographed on a 5 μM, 4.6 x 250 mm reversed phase Beckman Ultrasphere C18 analytical column (Beckman Coulter Inc., Fullerton, Calif.). The column was eluted isocratically with 2% acetonitrile in 50 mM ammonium formate (pH 6.5) for 10 minutes and then with a gradient from 0 to 10% acetonitrile over 10 minutes, followed by 10 to 80% acetonitrile over 10 minutes, 80 to 0% acetonitrile over 5 minutes and for the next 10 minutes the initial buffer was eluted isocratically. The flow rate was 1 mL/min. The compounds were isolated and collected from the reaction mixture by column chromatography on a 5 μm, 10 x 250 mm ODS Beckman Ultrasphere preparative column (Beckman Coulter Inc., Fullerton, Calif.) using the same elution method described above at a flow rate of 2 mL/min.

Spectroscopic and spectrometric methods

The ^{1}H-NMR spectra of MX, 2'-dGuo and the ethenoformyldeoxyguanosine (εfdGuo) adduct were recorded at 30°C on a Varian INOVA 500 NMR spectrometer at 500 MHz. The samples were dissolved in DMSO- d_6 and TMS was used as internal standard. The ^{1}H-NMR signal assignments were based on H-H and C-H correlational data and chemical shifts. The UV spectra of the compounds were recorded with the diode-array detector, set at 360 nm, as the peaks eluted from the HPLC column. The electrospray ionization (ESI) mass spectra were recorded on a Finnigan LCQ^{Deca} (ThermoQuest, San Jose, Calif.) ion trap mass spectrometer equipped with an API2 electrospray ionization (ESI) source, a Surveyor HPLC system (ThermoQuest), and a Surveyor autosampler (ThermoQuest) with a refrigerated sample tray. Nitrogen was both the nebulizing and the bath gas during ionization with a sheath gas flow rate of 79.98 psi. A voltage potential of 5.56 kV and a current of 14.99 μA were applied to the ESI needle. The capillary temperature and voltage were 250 °C and 7.24 V, respectively. The flow rate of the 50% methanol water mobile phase was 100 μL/min. A 5 μL sample loop was used. Finnigan Xcalibur™ software was used for data analysis.

Prepare MX-deoxyguanosine reaction mixture

The ethenoformyl adduct was synthesized according to the method published by Munter *et al* (1999). MX (358 mg, 1.65 mmol) was reacted with 2'-deoxyguonosine (220 mg, 0.82 mmol) in 145 mL of 0.01M phosphate buffer solution (pH 8.2). The pH of the reaction was checked daily and readjusted with 1M NaOH. The reaction was performed at 37°C for 7 days and at completion the reaction mixture was cooled overnight at 4°C. The mixture was filtered and the black precipitate (insoluble in water and DMSO) was discarded. Water was removed from the mixture via lyophilization and a brown solid was obtained. Two hundred and twenty (220 mg) milligram of the powder was dissolved in 5 mL ddH$_2$O. Analytical HPLC using reversed phase C18 column separated the εfdGuo adduct from the unreacted starting materials and other impurities. The fraction of interest with a UV-visual spectrum, similar to that obtained by Munter *et al.*, 1999, eluted at 15.7 minutes. The mixture was desalted using the preparative C18 column and fractions of interest were collected. The desalted fractions were combined and evaporated to dryness via lyophilization and the residue (3 mg dissolved in ddH$_2$O) was subjected to spectroscopic and spectrometric studies. About five hundred (500 μg) microgram of the solute was dissolved in 1mL of ddH$_2$O for liquid chromatography with electrospray ionization tandem mass spectrometry (LC-ESI/MS). The flow rate of the mobile phase (50% methanol water) was 100 μL/min and the sample volume per injection was 5 μL.

^{32}P-POSTLABELING ANALYSIS OF 1,*N*^{6}-ETHENODEOXYADENOSINE AND 3,*N*^{4}-ETHENODEOXYCYTIDINE

DNA was isolated and digested enzymatically to deoxyribonucleoside 3'-phosphates by using a micrococcal nuclease and spleen phosphodiesterase. The etheno deoxyribonucleotides were isolated from the normal nucleotides by immunoaffinity chromatography by using microcolumns that contain specific monoclonal antibodies (Misra *et al.*, 1994). The isolated adducted nucleotides were 5'-end labeled with ^{32}P-ATP in a reaction catalyzed by polynucleotide kinase. The labeled nucleotides were resolved by two-dimensional thin layer chromatography by using PEI-cellulose plates. Adducts were visualized and quantitated by autoradiography and liquid scintillation counting, respectively. DNA adduct levels were calculated using standard curves generated from known amounts of etheno modified deoxyribonucleotides.

ANALYSIS OF DNA ADDUCTS BY HPLC ELECTROSPRAY IONIZATION MASS SPECTROMETRY

HPLC-ESI/MS/MS analyses were conducted by using a Micromass Quatro II LC-ES MS/MS triple quadrupole system interfaced with an HP 1050 liquid chromatograph. Separation was evaluated by using a 3 μm C18 Hypersil column (Phenomenex, Torrance, Calif.) at a flow rate of 150 μL/min. Two mobile phases were used: (A) 5 mM ammonium formate, pH 3.5 and (B) 70% methanol in a linear gradient from 15 to 48% B over 10 min.

Nitrogen was used as drying gas. The source temperature was held at 120 °C and the cone energy at 35 V. Collision induced dissociation was achieved by using argon at 2.5×10^{-3} mb in Q2 at a collision energy of 20 eV. Parent ion scanning was performed by setting the parameters in quadropole Q3 to allow the protonated ions of a selected base and scan Q1 in the range m/z = 100-600 to detect the parent ions, giving rise to this fragment. This enabled selective detection of putative adducts. Once identified, selective ion monitoring was used to quantitate the adduct and its internal standard.

ANALYSIS OF GLUTATHIONE BY HPLC-ELECTROCHEMICAL DETECTION

One hundred milligram (100 mg) tissue was homogenized and an aliquot was removed for protein quantitation (BCA Protein Assay, Pierce Biotechnology, Rockford, Ill.). The remainder of the homogenate was precipitated by using trichloroacetic acid followed by centrifugation (Carro-Ciampi *et al.*, 1988). The supernatant was filtered and analyzed by HPLC-ECD. The electrochemical potential settings were set at regular intervals to 1100 mV. Quantitation on the response was compared to that of a known amount of methionine, which was added as an internal standard.

STATISTICS

The Student's t-test and analysis of variance (ANOVA) procedures were used to determine statistical significance among groups. In studies involving multiple measurements over time, repeated measures analysis of variance and regression methods were used.

CHAPTER 3: RESEARCH OVERVIEW

ORIGINAL RESEARCH PLAN - SPECIFIC AIM 1.

Identify DNA Adducts Derived From Glutathione Conjugates of Bromodichloromethane, and Quantitate Their Concentrations in Tissues of Exposed Rats and Mice—E. Bodes, H. Koc, R. Sangaiah, P. Begemann, D.K. La and James A. Swenberg

Overview

The conjugation of xenobiotics with glutathione is generally an important mechanism in cellular detoxification but the significance of glutathione in mediating the toxicity and carcinogenicity of BDCM is unclear. Gao *et al* (1996) reported that glutathione protects against the toxicity and macromolecular binding of BDCM. The protective effects were similar to those observed previously in studies of chloroform, where glutathione depletion increased hepatic and renal toxicity and increased covalent binding to hepatic proteins (Stevens and Anders, 1981). However, Pegram *et al* (1997) reported that BDCM-induced genotoxicity is derived from its conjugation with glutathione. Increased mutations occurred in *Salmonella typhimurium* transfected with glutathione *S*-transferase θ. This finding was consistent with that of Thier *et al* (1993), who showed that the expression of glutathione *S*-transferase θ was important for the mutagenicity of dihalomethanes. Glutathione conjugation also appeared to be associated with the induction of liver and lung cancer in mice by dihalomethanes (Anderson *et al.*, 1991). Pegram *et al* (1997) indicated that glutathione *S*-transferase-mediated transformation of BDCM to mutagenic intermediates occurs at low substrate concentrations and suggested potential risk to exposed human populations by this mechanism.

We hypothesized that DNA adduct formation by BDCM derives from its conjugation with glutathione. The identification of BDCM-induced DNA adducts would indicate whether glutathione is involved in its activation. We proposed to measure the extent of adduct formation in exposed rat liver tissues to study the relationship between their DNA adduct concentrations and susceptibility to carcinogenesis.

Glutathione *S*-transferase-mediated biotransformation of BDCM to reactive intermediates likely involves the formation of *S*-(1-dihalomethyl)glutathione (Pegram *et al.*, 1997). We attempted to synthesize this glutathione conjugate using a method developed for the synthesis of conjugates derived from haloethanes (Humphreys *et al.*, 1990). In follow-up experiments we decided to react the synthesized glutathione conjugate with calf thymus DNA to produce adducts with subsequent removal from DNA by neutral thermal, mild acid, or enzymatic hydrolysis followed by HPLC purification. (Beranek *et al.*, 1980). After characterization of the structures of these DNA adducts by mass spectrometry and NMR spectroscopy, we hoped to use these adducts as standards for subsequent studies *in vivo*.

DNA adduct formation by BDCM was investigated *in vivo* by exposing F344 rats to a single gavage dose of 0, 0.3, 1.0, or 3.0 mmol/kg (0, 49.17, 163.9 or 491.7 mg/kg) with five animals per dose group. Animals were sacrificed 6 hours post-exposure and different target and non-target organs were removed for analysis of DNA adducts. The method of DNA hydrolysis depended on the type of adducts that were formed. We planned to release adducts of guanine and adenine by neutral thermal or mild acid hydrolysis, whereas enzymatic digestion was the method of choice for adducts of pyrimidines (Beranek *et al.*, 1980). DNA adducts will be purified by HPLC, desalted by C18 solid phase chromatography, and analyzed by HPLC ESI/MS. Quantitation would be based on comparisons with internal standards (stable isotope derivatives).

We have used a similar approach previously to synthesize and characterize DNA adducts of the environmental contaminant, 1,2,3-trichloropropane (1,2,3-TCP; La *et al.*, 1995). We identified the major DNA adduct of 1,2,3-TCP as *S*-[1-(hydroxymethyl)-2-(*N*7-guanyl)ethyl]-glutathione. This DNA adduct derives from glutathione conjugation of 1,2,3-TCP by a mechanism similar to that proposed for BDCM.

Dechert and Dekant (1996) identified *S*-[1-N^2-deoxyguanosinyl)-methyl]glutathione from reaction of the dichloromethane metabolite, *S*-chloromethylglutathione, with deoxyribonucleosides. We expected *S*-(1-dihalomethyl)glutathione to induce similar DNA adducts. We also expected that the high sensitivity of HPLC ESI/MS would enable the detection of any other DNA adducts that are formed. It is possible that the genotoxicity induced by trihalomethanes may not result from direct binding to DNA, but rather through indirect mechanisms involving endogenous DNA adducts (Specific Aim 4). Glutathione *S*-transferase, while facilitating the conjugation of trihalomethanes with glutathione, may cause glutathione depletion, which is known to enhance concentrations of endogenous DNA adducts (Nath *et al.*, 1997). This alternative hypothesis is consistent with findings by Gao *et al* (1996) that glutathione protects against the toxicity and BDCM to cellular macromolecules.

Progress On Synthesis Of A Glutathione Conjugate

There is evidence to suggest that BDCM-induced genotoxicity derives from its conjugation with glutathione. Increased mutations occurred in *Salmonella typhimurium* transfected with glutathione *S*-transferase θ (GSTT1) (Pegram *et al.*, 1997). DeMarini *et al.* (1997) reported that the mutations induced by BDCM in GSTT1-transfected *Salmonella typhimurium* were predominantly GC to AT transitions.

Our objective was to synthesize the glutathione conjugate of BDCM using a method developed for the synthesis of haloethane-derived glutathione conjugates (Humphreys *et al.*, 1990). We planned to purify the conjugate using HPLC with UV diode array detection followed by characterization of the structure by MS and NMR spectroscopy. In addition, we planned to react this putative metabolite with DNA and individual deoxyribonucleosides to determine if BDCM is activated by conjugation with glutathione. We proposed to use the NMR characterized reaction products as standards in subsequent analyses.

Initial method used for conjugate synthesis. Glutathione (1 g) was dissolved in freshly prepared sodium methoxide (240 mg sodium added to 25 mL distilled methanol) and added drop-wise to a stirred solution of BDCM (32 mmol) in 25 mL of methanol. The solution was stirred for 1 hour. The glutathione conjugate of BDCM was precipitated with acetic acid, dried under a vacuum and stored at 4°C. We analyzed the product by HPLC with UV diode array detection and by liquid chromatography electrospray ionization mass spectrometry (HPLC-ESI/MS/MS).

We did not detect the dihalomethyl-glutathione conjugate after reacting BDCM with glutathione. HPLC with UV diode array analysis produced one major peak with the same retention time as glutathione. Analysis by MS did not indicate compounds other than glutathione, suggesting possible degradation of the product (Figure 3.1).

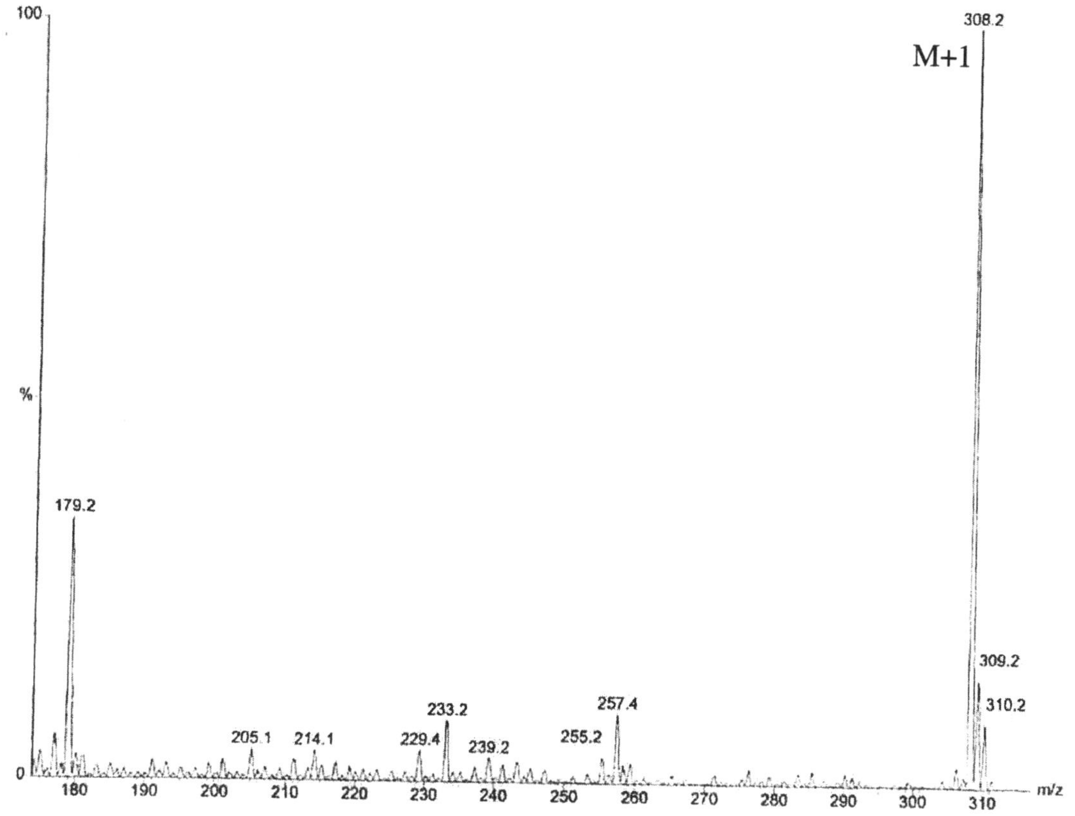

Figure 3.1. ESI/MS spectrum of glutathione prior to analyzing the reaction mixture (*m/z* = 308.2 = M+1). A 10 μM GSH solution was prepared in 50:50 water:acetonitrile solution. The elution buffer was 0.1 % formic acid at a flow rate of 0.5 μL/min.

Although the conjugate appears to be readily degradable, it may have sufficient half-life *in vivo* to react with macromolecules. Many other DNA reactive metabolites also are known to be highly labile. The stability of the dihalomethyl-glutathione conjugate is very different than that of the haloethane conjugates (Humphreys et al., 1990). Based on its structure, we expected that the conjugate formed between BDCM and glutathione to decompose rapidly to formyl chloride, formaldehyde, and glutathione. It may be possible that formyl chloride and formaldehyde are responsible for the observations cited originally as evidence of genotoxicity. However, mutation data suggests that formaldehyde is not likely to be responsible. Graves *et al* (1996) reported that formaldehyde induces transversions at AT base pairs and it also may participate in the production of DNA-protein cross-links (Dekant and Vamvakas, 1993). Based on this data, Graves and colleagues (1996) concluded that formaldehyde was not responsible for the mutations induced by methylene chloride, which induces GC to AT transitions. Instead, they suggested that S-chloromethylglutathione may play a role in methylene chloride mutagenesis.

Incubate glutathione with deoxyadenosine. We did not detect the BDCM glutathione conjugate, and speculated that the conjugate degraded during its isolation. We decided to incubate the reaction mixture directly with 0.32 mmol deoxyadenosine in 0.1 M Tris HCl (pH 7.4) to minimize possible degradation of the conjugate. The reaction mixture consisted of 25 mg glutathione added to 5 mL of a mixture containing 5 mL methanol and 25 mg sodium. Products of this reaction would be considered as potential adducts.

After incubation for 24 hours, the reaction mixture was precipitated in acetic acid, dried under vacuum and dissolved in 200 μL distilled deionized water for analysis by HPLC with UV diode array detection. Analytical HPLC was performed on a liquid chromatographic system consisting of two model 510 Waters HPLC pumps (Millipore), a 1040A-HPLC diode-array detector (UV; Hewlett-Packard), a Waters automated gradient control system and Hewlett Packard Chemstation for LC™ data analysis program. The reaction mixture was chromatographed on a 5 μm, 4.6 x 250 mm reversed phase Beckman Ultrasphere C18 analytical column (Beckman Coulter Inc., Fullerton, CA). The column was eluted isocratically with 5 % acetonitrile in 50 mM ammonium formate (pH 4.25) for 20 minutes. The flow rate was 1.5 mL/min.

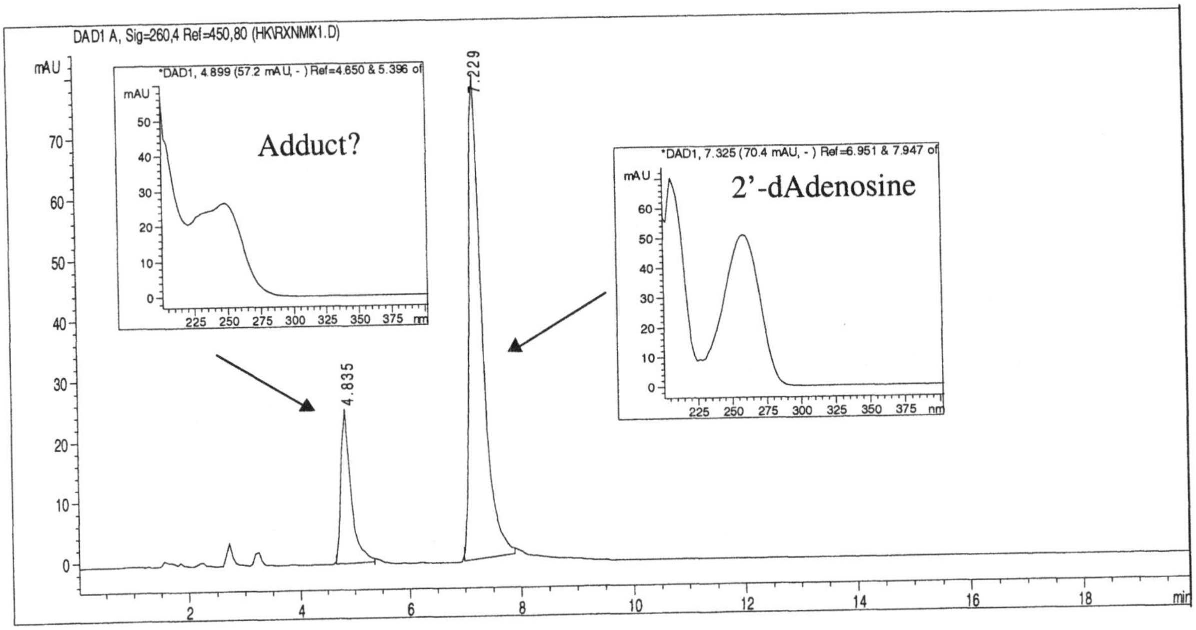

Figure 3.2. HPLC separation from a reaction mixture of BDCM, glutathione and 2'-dAdo with UV spectra of the suspected adduct and 2'-dAdo.

We detected one peak that eluted prior to adenosine and had a unique UV spectrum (Figure 3.2). This fraction was collected and analyzed by ESI/MS. The controls in this experiment were adenosine or the synthesis product (glutathione conjugate) in 0.1 M Tris-HCl. The reaction of the BDCM and glutathione with adenosine resulted in a product with UV and MS spectra that were suggestive of an adduct and the detected ions were different than those in control samples (Figure 3.3). However, the potential DNA adduct could not be further characterized because of the low yield. We were also unable to identify the possible chemical structure that was consistent with the mass spectrum. Nevertheless, this observation supports the idea that a reactive, short-lived compound such as *S*-(1-dihalomethyl)glutathione conjugate could be responsible for DNA adduct formation.

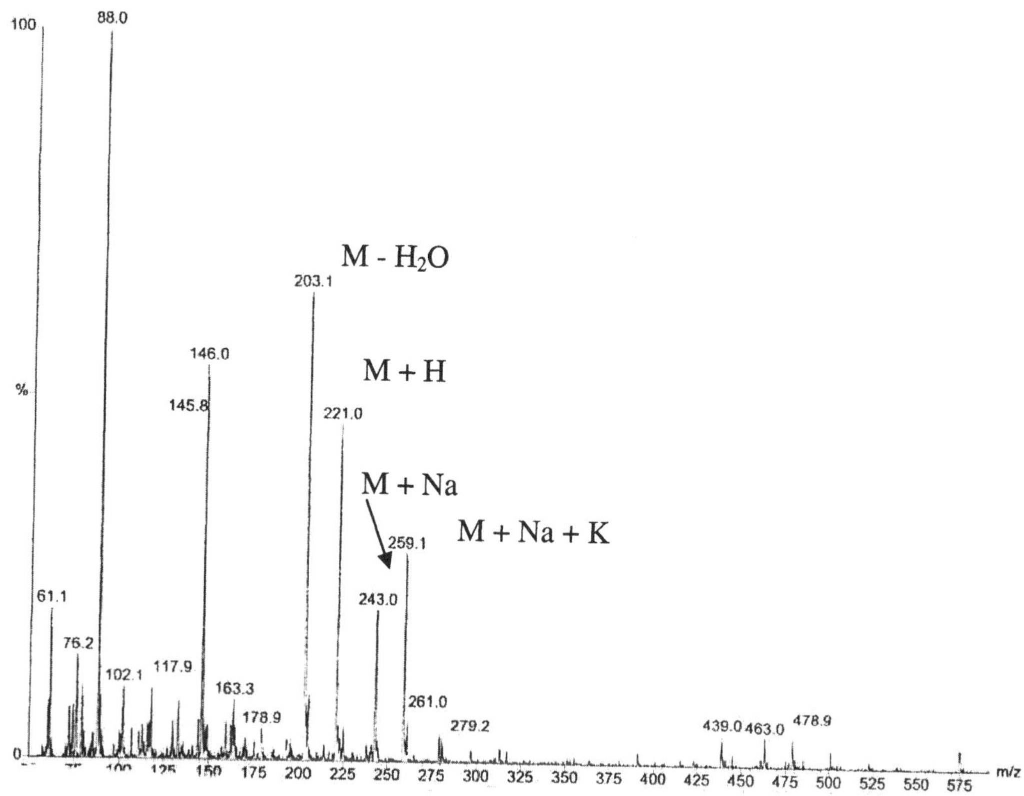

Figure 3.3. Mass spectrum of suspected adduct formed from the reaction between BDCM, glutathione and 2'-dAdo. The elution buffer was 0.1 % formic acid at a flow rate of 0.5 μL/min. The *m/z* results were as follows: 203.1, 221.0, 243.0 and 259.1 represented M - H_2O, M + H, M + Na and M + Na + K, respectively.

Conclusions

BDCM has been shown to be metabolized to intermediates that covalently bind to lipids and proteins and this binding has been associated with renal and hepatic toxicity (Gao *et al.*, 1996) in rodents. Consumption of trihalomethanes in chlorinated water has been implicated in human epidemiology studies in urinary bladder and intestinal tract cancers (Morris *et al.*, 1992). Evidence suggests that BDCM-induced genotoxicity is derived from its conjugation with glutathione (Pegram *et al.*, 1997). It is widely known that glutathione conjugate formation is an important bioactivation mechanism for several groups of compounds with implications for organ-selective toxicity and carcinogenicity. For example, vicinal dihaloalkanes such as 2,3-dihaloethanes yield conjugates that can give rise to highly electrophilic episulphonium ions that are involved in the cytotoxicity and mutagenicity of 1,2-dihaloethanes (Anders *et al.*, 1992; Figure 3.4). Initially it was thought that BDCM may cause direct-acting glutathione conjugates, like 1-bromo-2-chloroethane, that can form electrophilic episulphonium ions. However, this is not possible since BDCM contains only a single carbon atom. It is more likely that BDCM forms direct toxic glutathione *S*-conjugates like halogenated alkanes (e.g. dichloromethane) (Dekant and Vamvakas, 1993).

Our objective for this specific aim was to synthesize, purify and characterize the dihalomethyl-glutathione conjugate formed between BDCM and glutathione. After a complete characterization of the conjugate product by HPLC, MS and NMR we proposed to analyze DNA from exposed animals to determine whether this adduct(s) is also formed *in vivo*. Unfortunately we were not able to detect any conjugate. Both the HPLC and mass spectrometry data did not indicate compounds other than glutathione. This may suggest that the reaction product is labile which is quite different than that of haloethane-glutathione conjugates (Humphreys *et al.*, 1990).

The follow-up reaction between the BDCM-glutathione mixture and adenosine resulted in a product with UV and MS spectra that were suggestive of an adduct and the detected ions were different than those in control samples. However, the potential DNA adduct could not be further characterized because of the low yield in which it was produced. We were also unable to identify the possible chemical structure that was consistent with the MS data. Nevertheless, this observation supported the idea that a reactive, short-lived compound such as *S*-(1-dihalomethyl)glutathione conjugate could be responsible for DNA adduct formation.

Future Research Direction

Since the reaction between BDCM and glutathione with adenosine resulted in a product with UV and MS spectra that were suggestive of an adduct, this experiment should be repeated. Additional reactions with guanosine and calf thymus DNA should also be conducted. These reactions have to be scaled to yield products in quantities sufficient for mass spectrometry and NMR spectroscopy analyses for the identification of the chemical structure of the suggestive adduct product. Following the characterization of DNA adduct standards, DNA from target tissues of BDCM-exposed rats could be analyzed. However, at the combined meeting between our research group, the PAC and Project Manager on November 14, 2000 the group decided that we should not proceed with this specific aim but focus our efforts on the Objectives outlined under "Revised Research Plan".

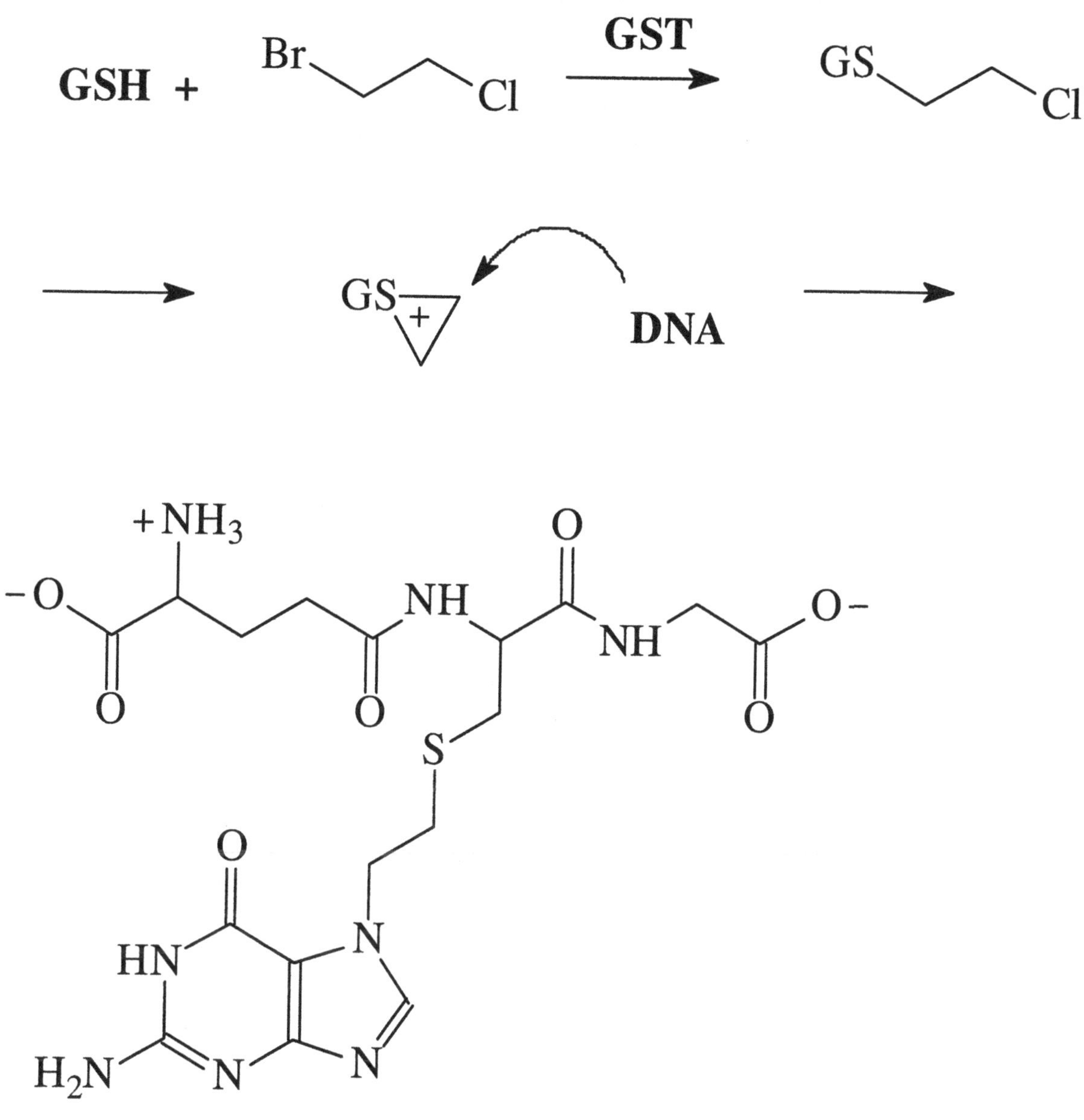

Figure 3.4. Episulphonium ion formation during glutathione-dependent bioactivation of 1-bomo-2-chloroethane to a mutagenic metabolite.

ORIGINAL RESEARCH PLAN - SPECIFIC AIM 2.

Identify And Quantitate DNA Adducts Derived From Mutagen X—E. Bodes, H. Koc, R. Sangaiah, A. Gold, D. K. La And J. A. Swenberg

Overview

3-Chloro-4-(dichloromethyl)-5-hydroxy-2(*5H*)-furanone (MX) is a potent, direct-acting mutagen that can account for as much as 60% of the mutagenic activity in chlorinated drinking water (Brunborg *et al.*, 1991; DeMarini *et al.*, 1995). The mechanism underlying the genotoxicity of MX in *Salmonella* is not fully understood, and little is known about the chemistry between MX and DNA. Because of its potent mutagenicity, there is a considerable interest in identifying DNA adducts of MX.

DNA adduct formation by MX has been studied previously by ^{32}P-postlabeling. Its reaction with calf thymus DNA yielded two DNA adducts that increased in a dose-dependent manner (Schut *et al.*, 1991). Concentrations up to 9.4 per 10^6 nucleotides were measured. Preliminary studies also indicated the formation of these adducts in B6C3F1 mice exposed orally to MX. Meier *et al* (1989) reported that a single DNA adduct increased with dose in bacterial and mammalian cells with adduct concentrations up to 2 per 10^7 were measured. While ^{32}P-postlabeling is very sensitive for DNA adduct detection and quantitation, this method does not provide any structural information about adducts. The choice of specific DNA adducts for use as biomarkers ideally should involve those that are biologically relevant. Identification of these DNA adducts would likely provide important information about their relevance to MX-induced mutagenesis and carcinogenesis.

Based on DNA sequencing, Marsteinstredet *et al* (1997) reported that DNA adducts of MX occurred preferentially at guanine. Alhonen-Raatesalmi and Hemminki (1991) attempted to identify DNA adducts of MX from its reaction with deoxyribonucleosides. They did not detect stable DNA adducts using HPLC with UV detection which suggested that the adducts were labile. However, the detection of DNA adducts by ^{32}P-postlabeling indicates that the MX-derived adducts are stable but could not be detected by HPLC with UV detection due to sensitivity. Munter *et al* (1998) identified two adenine adducts in calf thymus DNA following reaction with MX. The importance of these adenine adducts is unclear as the predominant mutations induced by MX are GC to TA transversions. We proposed to focus primarily on DNA adducts of deoxyguanosine, as mutational spectra indicate their involvement in mutagenesis and carcinogenesis.

The objective of this Specific Aim was to identify DNA adducts of MX using HPLC electrospray ionization tandem mass spectrometry (HPLC-ESI/MS/MS). Mass spectrometry has sufficient sensitivity to detect quantities reported by Schut *et al* (1991) and Meier *et al* (1989). Since MX is direct acting, we decided to incubate MX with calf thymus DNA as well as with individual deoxyribonucleosides. In addition, we decided to synthesize MX according to a published method by Franzen and Kronberg (1995) since our *in vitro* and *in vivo* experiments required large amounts of MX. We expected that HPLC-ESI/MS/MS would enable sensitive detection of MX-derived DNA adducts as well as provide important structural information.

Synthesis Of Mutagen X

Although MX is available commercially, our planned experiments (*in vitro* and *in vivo*) required large quantities of this compound. In collaboration with Dr. Ramaih Sangaiah (Dept. of Environmental Sciences and Engineering, University of North Carolina at Chapel Hill), we synthesized MX following a published method by Franzen and Kronberg (1995). The product was dissolved in CD_3OD (deuteriated methanol) and analyzed by ^1H-NMR. The NMR spectrum was consistent with the structure of MX, and its purity was calculated to be greater than 95%. The synthesized product was stored in ethyl acetate at –20 °C, i.e. conditions under which MX has been reported to be stable. We expected that MX was in its ring open form under these conditions. The ethyl acetate was evaporated under nitrogen, and dissolved in corn oil for animal exposure. MX was reconstituted in 0.5 M sodium phosphate (pH 8.2) for synthesis reactions, consistent with conditions reported by Munter *et al* (1999) in their synthesis of the ethenoformyl guanosine adduct.

Synthesis And Quantitation Of Ethenoformyldeoxyguanosine Adduct Derived From MX

MX is one of the most potent direct-acting mutagens tested in bacterial systems. Komulainen and colleagues (1997) reported that MX induced cholangiomas and thyriod tumors in rats exposed to MX via drinking water for 104 weeks. The objective of this Specific Aim was to identify DNA adducts of MX using HPLC electrospray ionization tandem mass spectrometry.

A recent publication by Munter *et al* (1999) reported the detection of an ethenoformyl adduct [3-(2'-Deoxy-β-D-ribofuranosyl)-7-formyl-imidazo[1,2-α]purin-9(4H)-one (εfdGuo)] following reaction of MX with deoxyguanosine. We proposed to repeat the method reported by Munter and co-workers to generate the ethenoformyl adduct as well as other DNA adducts that may be formed. We dissolved deoxyguanosine in 0.5 M sodium phosphate (pH 8.2) and the reaction mixture was added to MX. We used MX that we synthesized as well as MX purchased from Sigma Aldrich Chemical Co. (St. Louis, Mo.). Incubations were at 37 °C for up to four days. The reaction mixtures were analyzed by HPLC with UV diode array detection (Figure 3.5).

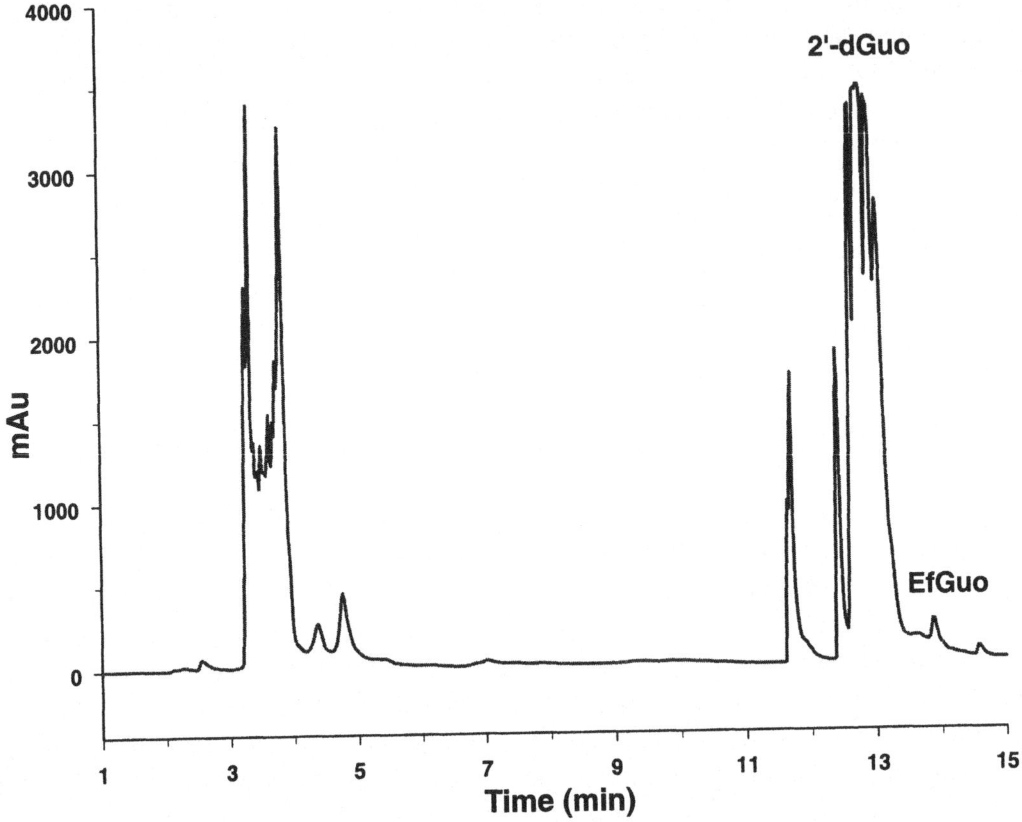

Figure 3.5. MX-2'-dGuo reaction mixture separated by HPLC.

Several peaks had UV spectra that were indicative of adducts (Figure 3.6). Fractions corresponding to peaks were collected, desalted by solid phase chromatography, and analyzed by LC/MS/MS. Mass spectrometry analysis indicated that εfdGuo was suitable for MS/MS detection in SRM mode since the M+1 ion resulting from protonation of εfdGuo efficiently produced a product ion at *m/z* 204 due to neutral loss of 116 atomic mass units (amu), i.e. deoxyribose. Adducts of deoxynucleosides generally fragment by loss of the deoxyribose, yielding adducts of the corresponding nucleobases. By this approach, we identified several ions that resulted in loss of deoxyribose (Figure 3.7). We then conducted MS/MS analysis to examine the fragmentation of parent ions. These ions appeared to be precursors of the ethenoformyl adduct that was identified by Munter *et al* (1999).

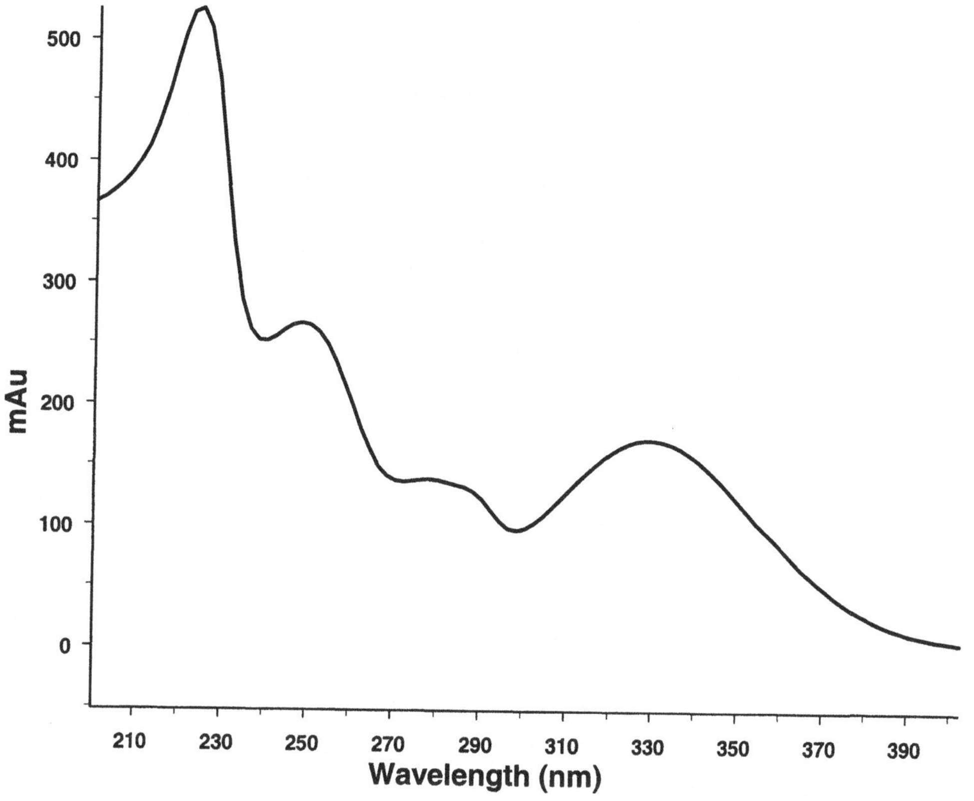

Figure 3.6. **UV spectrum of suspected εfdGuo adduct with UV_{max} values at 230 and 330 nm.**

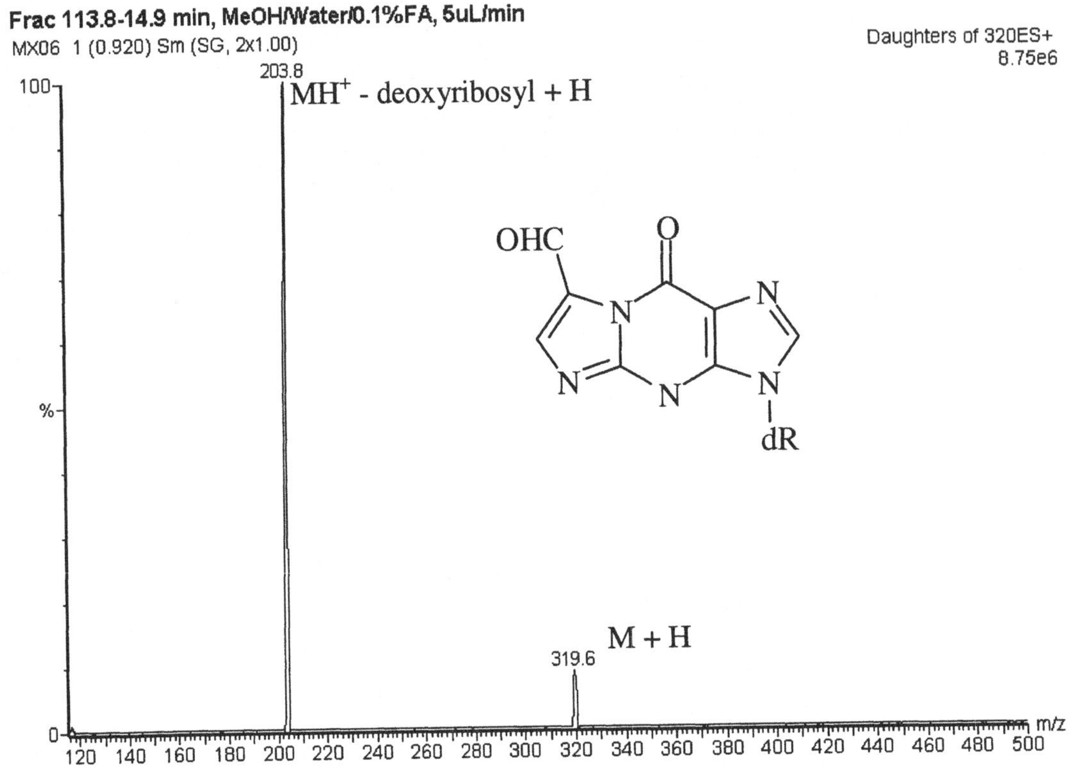

Figure 3.7. Representative scan from HPLC-ESI/MS analyses of the reaction mixture of MX and 2'-dGuo. The scans were performed to detect ions that give rise to a loss of mass 116, corresponding to the deoxyribose.

Conclusions

There is considerable interest in identifying DNA adducts of MX because of its potent mutagenicity and the utility of MX derived adducts as biomakers of exposure. Munter et al., 1999 reported that an ethenoformyl adduct [3-(2'-Deoxy-β-D-ribofuranosyl)-7-formyl-imidazo[1,2-α]purin-9(4H)-one] formed in a reaction of MX with deoxyguanosine (1,N^2-ethenoformyldeoxyguanosine; εfdGuo). However, their research group was not able to detect the εfGuo adduct in calf thymus DNA and speculated that the adduct may be formed at levels lower than the detection limit of their HPLC analytical method (~5 adducts /10^7 bases).

Our research group focused on developing a method where large amounts of the εfdGuo adduct can be formed when MX is reacted with 2'-deoxyguanosine. Initially we synthesized MX in collaboration with Dr. Ramiah Sangaiah (Dept. of Environmental Sciences and Engineering, Univ. of North Carolina at Chapel Hill) since we needed large amount of MX for our experiments. The ^{1}H-NMR spectrum was consisted with the structure of MX and its purity was calculated to be greater than 95%. We used MX that we synthesized as well as MX purchased from Sigma Aldrich Chemical Co. in the initial reactions. Preliminary analysis of reaction mixtures by HPLC with UV diode array detection suggested that several peaks had UV spectra that were indicative of adducts. We collected fractions corresponding to the peaks for analysis by LC/ESI/MS/MS. Mass spectrometry analysis indicated that the M+1 ion resulting from protonation of εfdGuo efficiently produced a product ion at m/z 204 due to neutral loss of deoxyribose (116 amu). After these promising results additional reaction mixtures were made and analyzed. This additional research is discussed under the "Revised Research Plan – Specific Aim 9".

ORIGINAL RESEARCH PLAN - SPECIFIC AIM 3.

Quantitate Concentrations of Etheno DNA Adducts in Rats Exposed to Mucochloric Acid (Mca)—E. Bodes, E. J. Morinello, D. K. La and J. A. Swenberg

Overview

DNA adducts derived from 3,4-dichloro-5-hydroxy-2(*5H*)-furanone (MCA) have been identified previously. *In vitro* reaction of MCA with nucleosides results in the formation of etheno and ethenocarbaldehyde derivatives (Kronberg *et al.*, 1992, 1993), however, these adducts have not been identified *in vivo*. Etheno DNA adducts are promutagenic lesions that are induced by a number of carcinogens, including vinyl chloride and urethane (La and Swenberg, 1997; Figure 3.8).

1,N^2-εdeoxyguanosine N^2,3-εdeoxyguanosine   1,N^6-εdeoxyadenosine

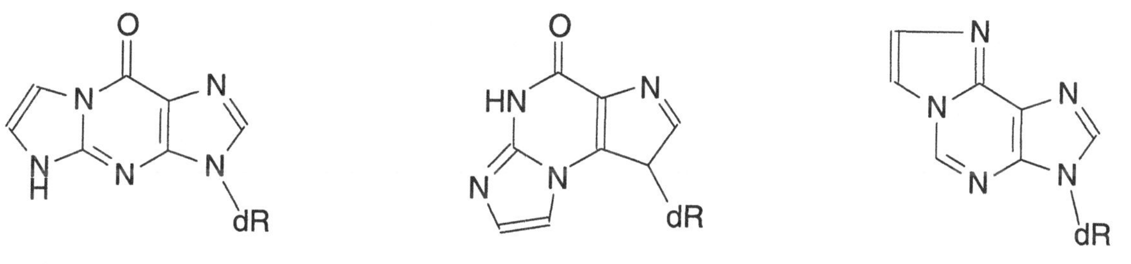

Figure 3.8. Structures of etheno DNA adducts.

Etheno adducts also result from the lipid peroxidation product, *trans*-4-hydroxy-2-nonenal, and are present in DNA endogenously (La and Swenberg, 1997). Consequently, etheno DNA adducts in tissues of MCA-exposed animals may derive from MCA or lipid peroxidation. Since MCA had not been investigated previously *in vivo*, we decided to expose F344 rats to 100 mg/kg. Animals were sacrificed 4, 12 or 24 hours post-exposure, and the liver was removed for DNA isolation and adduct analysis by GC/MS (Detailed Materials and Methods). We have developed an ultrasensitive and ultraspecific gas chromatography mass spectrometry method to detect N^2,3-εGua adducts. We planned to develop similar mass spectrometry methods for analyses of 1,N^2-εGua and 1,N^6-εdA.

We expected that gas chromatography mass spectrometry would be a powerful tool for the analysis of MCA-induced DNA adducts, not only in terms of its sensitivity, but also its specificity, which would enable us to distinguish between endogenous and exogenously-derived DNA adducts.

To determine the source of etheno DNA adducts in MCA-exposed animals, we proposed initially to synthesize a stable isotope [^{13}C]-derivative of MCA (Franzen and Kronberg, 1995), and to use mass spectrometry to differentiate DNA adducts derived from MCA and lipid peroxidation. Dr. Ramiah Sangaiah has designed a synthesis method for [^{13}C]-MCA from [^{13}C]-deoxyribose.

Develop Methods To Quantitate Etheno Adducts

Although MCA is not as abundant in drinking water as trihalomethanes and haloacetic acids, it has several interesting features that merit its investigation. First, MCA and other halogenated furanones are potent mutagens. In addition, several of the DNA adducts induced by MCA are also formed endogenously from lipid peroxidation. DNA adduct formation by MCA *in vitro* is known, as Kronberg *et al* (1992 and 1993) have reported that the reaction of MCA with nucleosides results in etheno and ethenocarbaldehyde derivatives. We have developed an assay for the detection and quantitation of an etheno DNA adduct, N^2,3-εGua (Ham *et al.,* 1999), using immunoaffinity (IA) chromatography coupled with gas chromatography/high resolution mass spectrometry (GC/HRMS). We have made significant improvements to the original method, for example, by including a neutral thermal method of depurination compared to acid hydrolysis. Neutral thermal hydrolysis results in selective depurination of *N*3 and *N*7 adducts of deoxyadenosine and deoxyguanosine, whereas acid hydrolysis results in complete depurination of deoxyadenosine and deoxyguanosine. We showed that N^2,3-εGua was released by neutral thermal hydrolysis providing a partial clean-up of the adduct. We also changed the method of DNA adduct purification by using immunoaffinity chromatography rather than cation exchange chromatography. The immunoaffinity chromatography columns were made using polyclonal antibodies developed specifically against N^2,3-εGua. We derivatized N^2,3-εGua using pentafluorobenzyl bromide and analyzed the derivative by GC/MS/MS, as previously described in the grant proposal. These changes have significantly improved the specificity and sensitivity of detection. With these changes a detection limit of 200 attomoles of standard has been achieved.

This method was adapted by Morinello *et al* (2001) to be used for the simultaneous quantitation of N^2,3-εGua and 1,N^2-εGua (Figure 3.9; Detailed Materials and Methods). Similarly, a fast and highly specific analytical method is being developed in our laboratory to quantify 1,N^6-εdA by using IA chromatography combined with liquid chromatography/ electrospray ionization mass spectrometry (LC/ESI-MS; Detailed Methods).

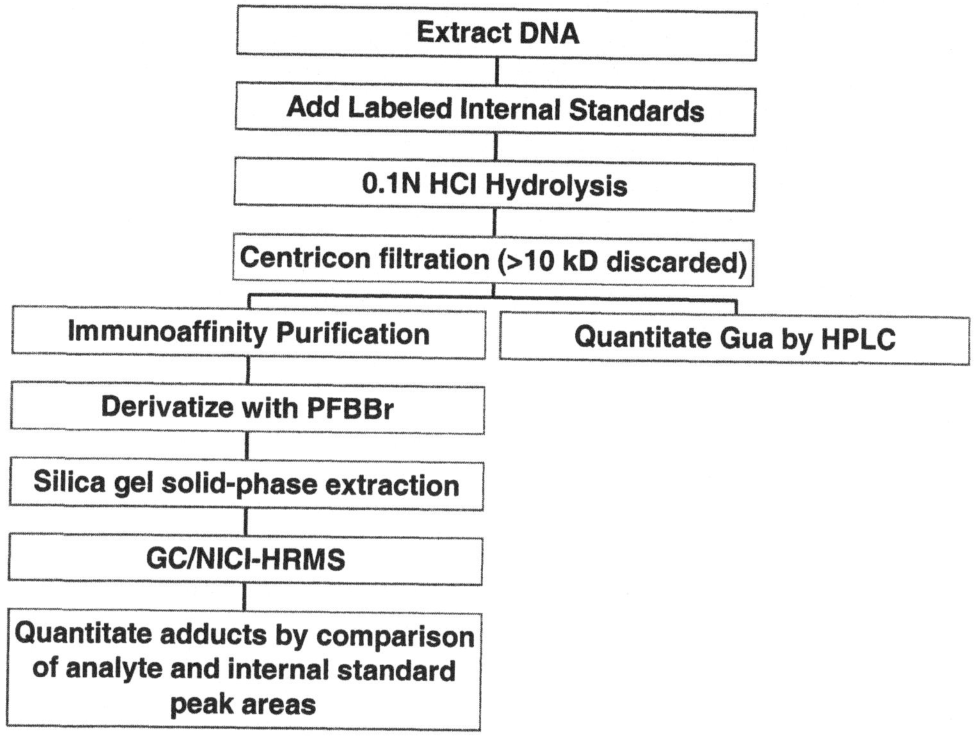

Figure 3.9. Method for the simultaneous quantitation of 1,N^2-ethenoguanine and N^2,3-ethenoguanine adducts.

The remainder of this Specific Aim is discussed under the "Revised Research Plan – Specific Aim 10" since most of that work was conducted after Dr. David La left the University of North Carolina.

ORIGINAL RESEARCH PLAN - SPECIFIC AIM 4.

Investigate Whether Disinfection By-Products Enhance Concentrations of Endogenous DNA Adducts—E. Bodes, B. F. Pachkowski, D. Lemaire, D. K. La and J.A. Swenberg

Overview

While considerable information exists on DNA adducts derived from exogenous sources; research has only recently focused on modifications produced by electrophiles that are generated endogenously. Endogenous DNA adducts, which include those derived from reactive oxygen species and lipid hydroperoxides, are present in DNA at levels that can exceed those derived from exogenous DNA adducts by 1000-fold. Increases in endogenous DNA adducts can arise by different means. One important mechanism may involve oxidative stress following the depletion of cellular defenses.

High-dose gavage exposure to chemicals (i.e. similar levels to those used in chronic animal bioassays) can deplete critical cellular defenses, such as glutathione, which can then result in increased oxidative stress and lipid peroxidation. For example, in studies of chloroform, glutathione depletion increased hepatic and renal toxicity (Stevens and Anders, 1981). Kluwe and Hook (1981) speculated that exposure to BDCM at doses used in chronic animal studies also depleted glutathione. Gavage administration of DCA and TCA to mice and rats produced significant increases in lipid peroxidation and 8-OHdG (Austin *et al.*, 1995, 1996). Increases were thought to reflect decreased activity of protective mechanisms. We extended the studies of Austin *et al* to include trihalomethanes like BDCM. Austin *et al* (1996) reported that maximum formation of thiobarbituric reactive substances, indicative of lipid peroxidation, occurs 3 hours following acute exposure to DCA. Maximum increases in 8-OHdG appear to occur 6 to 12 hours following exposure. We hypothesized that gavage exposure to these disinfection by-products will enhance endogenous DNA adducts secondary to depletion of cellular defense mechanisms.

There are many DNA adducts associated with endogenous processes. Rather than identifying and measuring a large number of endogenous DNA adducts, we decided to limit analyses to several important adducts. The hydroxyl radical-derived DNA adduct, 8-OHdG, as an indicator of damage from oxidative stress, and the etheno DNA adducts, 1,N^6-εdA, 3,N^4-εdC, and N^2,3-εGua, were measured as indicators of damage from lipid peroxidation.

We planned to focus initially on the liver of exposed rats and mice for DNA isolation and analysis to quantitate endogenous DNA adducts. 8-OHdG was quantitated by HPLC-ECD, εdA and εdC were analyzed by ^{32}P-postlabeling, and εGua by immunoaffinity chromatography coupled with gas chromatography/high resolution mass spectrometry (GC/HRMS) (Detailed Materials and Methods). We also planned to analyze tissues for glutathione by HPLC-ECD (Detailed Materials and Methods). We expected that high-dose gavage administration of drinking water disinfectants would enhance concentrations of endogenous DNA adducts.

Identify And Quantitate Adducts In Liver DNA From Rats Exposed To BDCM

We hypothesized that high-dose gavage exposure to chemicals would deplete critical cellular defenses, such as glutathione, which can result in increased oxidative stress and lipid peroxidation. These effects are not likely to occur when exposure involves lower doses or other routes of exposure. High-dose exposure, similar to those used in chronic animal bioassays, far exceed those present in drinking water and are not expected to be relevant doses for human risk assessment. These high-dose exposures are necessary to test whether enhancement of endogenous DNA adducts occur only at these concentrations and not at levels present in drinking water.

Single bolus gavage exposure to BDCM. We exposed male F344 rats to BDCM (0, 0.3, 1, or 3 mmol/kg, n=5 per group) by corn oil gavage. Rats were sacrificed 6 hours following a single exposure. This method of compound administration was used to maximize the likelihood of generating DNA adducts in vivo. In addition, the selected dose range was higher than that found in drinking water to ensure the detection of a discernable change in 8-OHdG, since oxidative stress generally occurs following high-dose exposures. However, if there was no change in 8-OHdG at these elevated levels we could conclude that oxidative stress mechanisms are not likely to be important at lower concentrations.

DNA from liver tissues was hydrolyzed and analyzed by HPLC-ECD for 8-OHdG. There was no significant difference between the control and treated groups (Table 3.1). We also determined the effect of BDCM on AP site formation in liver DNA (Specific Aim 8). We observed no significant increase in AP sites for all exposure groups. Our findings are consistent with current literature that does not support an induction in any DNA adducts or AP sites after exposure to BDCM. However, it is known that BDCM is genotoxic but the exact mechanism by which BDCM induces carcinogenesis is unknown. In addition, there is evidence to suggest that BDCM-induced genotoxicity derives from its conjugation with glutathione. However, since we were not able to identify a BDCM-glutathione conjugate (Specific Aim 1) it was not possible to determine concentrations of suspected adducts in exposed rats as proposed in the original research.

Table 3.1
8-OHdG in hepatic DNA of F344 rats exposed to BDCM.

Dose (mmol/kg)	8-OHdG (per 10^6 dG)
0	4.6 ± 0.2
0.3	6.7 ± 0.9
1	3.4 ± 1.7
3	3.1 ± 0.8

F344 rats received a single gavage dose and were sacrificed 6 hours later. 8-OHdG analysis was conducted by HPLC-ECD. Mean ± SD with $p > 0.05$

Identify And Quantitate Adducts In Liver DNA From Rats And Mice Exposed To Dichloroacetate (DCA)

Drinking water exposure to DCA for 28 days. In collaboration with Dr. Anthony DeAngelo (US EPA, Research Triangle Park, NC), Fischer-344 rats and B6C3F1 mice were exposed to DCA by drinking water for 28 days. Concentrations of DCA were 0, 0.5 or 1.5 g/L for rats, and 0 or 3.5 g/L for mice. We analyzed hepatic DNA for the hydroxyl radical derived DNA lesion, 8-OHdG, by HPLC-ECD. We also examined DNA for the lipid peroxidation induced DNA adducts, $1,N^6$-εdA and $3,N^4$-εdC by immunoaffinity chromatography enriched ^{32}P-postlabeling. We did not observe an increase in these endogenous DNA lesions (Tables 3.2 and 3.3). There appeared to be a decrease in etheno DNA adducts with exposure to DCA, but these differences were not statistically significant.

Table 3.2
Endogenous DNA lesions in F-344 rats exposed to DCA.

Concentration (g/L)	8-OHdG (per 10^6 dG)	εdA (per 10^8 dA)	εdC (per 10^8 dC)
0	2.7 ± 3.1	29.5 ± 15.8	31.2 ± 15.8
0.5	4.2 ± 5.8	12.7 ± 9.5	21.5 ± 14.7
1.5	2.8 ± 3.3	17.7 ± 11.9	20.0 ± 12.6

F344 rats were administered DCA (0, 0.5 or 1.5 g/L) in drinking water for 28 days. 8-OHdG analysis was conducted by HPLC-ECD. εdA and εdC quantitation were conducted by ^{32}P-postlabeling. Mean ± SD with $p > 0.05$.

Table 3.3.
Endogenous DNA lesions in B6C3F1 mice exposed to DCA.

	8-OHdG (per 10^6 dG)	εdA (per 10^8 dA)	εdC (per 10^8 dC)
Control	3.5 ± 0.5	5.4 ± 3.4	2.2 ± 1.2
Exposed	3.6 ± 0.4	2.5 ± 1.7	2.4 ± 1.3

B6C3F1 mice were administered DCA (0 or 3.5 g/L) in drinking water for 28 days. 8-OHdG analysis was conducted by HPLC-ECD. εdA and εdC quantitation were conducted by ^{32}P-postlabeling. Mean ± SD with $p > 0.05$.

In addition, we measured AP sites in hepatic DNA of the exposed mice (Specific Aim 8). There was no effect of exposure on AP site concentrations in the control group versus the exposed group. In collaboration with Dr. Steven Swarts (Department of Radiation Oncology, Wake Forest University), we also performed GC/MS analysis for 4,6-diamino-5-formamidopyrimidine (Fapy-adenine) and thymine glycol in control and DCA-exposed rats. We did not observe statistically significant changes in these endogenous DNA lesions.

Single bolus gavage exposure to DCA. Additional rats and mice were exposed by gavage to determine whether bolus administration increases endogenous DNA adduct formation. Rats were exposed to a single gavage administration of DCA at 100 mg/kg, and mice to 400 mg/kg. Animals were sacrificed 4, 8, or 24 hours following exposure and DNA was isolated from the liver for analysis.

Table 3.4.
8-OHdG in liver DNA of mice acutely exposed to 400 mg/kg DCA by gavage

	8-OHdG per 10^6 dG	
Time (h)	Control	Exposed
4	6.0 ± 1.2	5.6 ± 0.4
8	6.4 ± 1.2	6.4 ± 3.2
24	6.4 ± 0.4	8.4 ± 1.2

B6C3F1 mice were administered 400 mg/kg DCA by gavage and sacrificed 4, 8 or 24 hours later. 8-OHdG analysis was conducted by HPLC-ECD. Mean ± SD with $p > 0.05$.

We measured 8-OHdG by HPLC-ECD, N^2,3-εGua by gas chromatography/mass spectrometry (GC/MS), and 1,N^6-εdA and 3,N^4-εdC by ^{32}P-postlabeling. We believed that these experiments would provide information on whether endogenous DNA adducts are increased by high dose exposure to a disinfection by-product. In addition, the experiments would also indicate when maximum increases of endogenous DNA adducts occur following exposure, which would be important as we planned experiments to test other disinfection by-products for their effects on endogenous DNA lesions.

We did not observe an increase in the concentration of 8-OHdG in mouse liver DNA (Table 3.4). Previous studies have reported either small or no increase in 8-OHdG or lipid peroxidation from DCA exposure, depending on the dose and exposure method (Austin *et al.*, 1995, 1996; Parrish *et al.*, 1996). Larger increases have been reported for brominated acetates.

Conclusions

High-dose exposure to chemicals can result in oxidative stress and lipid peroxidation by depleting critical cellular defenses. We have examined whether high-dose exposure to BDCM or DCA affects DNA lesions associated with oxidative stress and lipid peroxidation. Male F344 rats and B6C3F1 mice were exposed to DCA via drinking water for 28 days at 0, 0.5 or 1.5 g/L for rats, and 0 or 3.5 g/L for mice. We analyzed hepatic DNA for 8-OHdG by HPLC-ECD. We also examined DNA for the lipid peroxidation induced DNA adducts, 1,N^6-εdA and 3,N^4-εdC by ^{32}P-postlabeling. We did not observe an increase in these endogenous DNA lesions. In addition, we measured AP sites in hepatic DNA of the exposed mice. We did not observe any significant difference between the control and exposed groups. We also exposed male F344 rats to BDCM (0, 0.3, 1, or 3 mmol/kg, n=5 per group) by corn oil gavage. Rats were sacrificed 6 hours following a single exposure. Hepatic DNA was hydrolyzed enzymatically to deoxyribonucleosides and analyzed for 8-OHdG by HPLC-ECD. We did not detect an increase in 8-OHdG following exposure to BDCM (Table 3.1). There appears to be a decrease in 8-OHdG following exposure to BDCM at the 1 and 3 mmol/kg dose but the difference was not statistically significant. In addition, there was no difference in AP site accumulation in any of the dose groups when compared to the control group (Specific Aim 8).

Our findings are consistent with current literature that does not support an induction in any DNA adducts or AP sites after exposure to BDCM. However, it is known that BDCM is genotoxic but the exact mechanism by which BDCM induces carcinogenesis is unknown. The doses selected for this study were similar to those used by Dunnick *et al* (1987) in a two-year gavage assay. The researchers found that the incidence of renal tubular cell adenomas and adenocarcinomas, and large intestinal adenomatous polys and adenocarcinomas were increased in both sexes of F344 rats. However, when administered in drinking water BDCM induced neoplastic nodules and adenofibrosis in F344 rat livers (IARC, 1991). In addition, there is evidence to suggest that BDCM-induced genotoxicity derives from its conjugation with glutathione. However, since we were not able to identify a BDCM-glutathione conjugate (Specific Aim 1) it was not possible to determine concentrations of suspected adducts in exposed rats as proposed in the original research.

The mean values in many instances are suggestive of some cause and effect relationships, but variance in the data sets perhaps prevents the ability to make salient observations. There are large variances in the 1,N^6-εdA and 3,N^4-εdC results which are inherent to the ^{32}P-postlabeling method and repeated measurements are unlikely to overcome this problem. Other approaches, such as mass spectrometry, which yield results with greater reproducibility are likely to reduce the variance and increase the likelihood of achieving statistical significance. We currently have a GC/MS method for the quantitation of N^2,3-εGua and 1,N^2-εGua, and a LC/MS method for the quantitation of 1,N^6-εdA. However, no additional liver tissue or DNA were available for determining adduct levels by mass spectrometry and the PAC advised us to spend more time on the other specific aims in this research plan.

ORIGINAL RESEARCH PLAN - SPECIFIC AIM 5.

Examine the Effects of Different Exposure Routes on Endogenous DNA Adducts—E. Bodes, B. F. Pachkowski, D. Lemaire, D. K. La and J.A. Swenberg

Overview

Many chemicals, including several disinfection by-products, are carcinogenic in experimental animals when administered by gavage. When animals are exposed by other routes, such as drinking water, many of these chemicals are not carcinogenic, or induce tumors at significantly lower incidences (La *et al.*, 1996). There is considerable debate over the relevance of high-dose gavage studies to human health, since human exposure to chemicals differs markedly from the mouse and rat exposed by gavage. Potential human exposures to disinfection by-products would occur by drinking water at lower concentrations. Based on consumption of 2 liters of water, potential human exposure would range from 0.1 to 0.3 μg/kg, several orders of magnitude lower than exposures in animal studies.

Gavage administration of chemicals results in bolus doses, whereas other exposure routes involve delivery over an extended period of time, which permits more efficient detoxication and elimination. Gavage administration is likely to produce greater concentrations of test substances at target sites. This difference can affect saturable processes, such as bioactivation, detoxication, and DNA repair, resulting in important quantitative differences in toxicity and carcinogenicity. One area of difference between gavage and drinking water exposure to chemicals may be in their effects on endogenous DNA adducts. Bolus exposure to chemicals is likely to deplete cellular defenses, and enhance concentrations of endogenous DNA adducts derived from reactive oxygen species and lipid peroxidation. One would not expect a significant increase in endogenous DNA adducts to occur with drinking water exposure.

The objective of this specific aim is to investigate whether exposure route affects endogenous DNA adducts. We expected that gavage administration would increase endogenous DNA adducts, whereas drinking water exposure would not have any effects on their concentrations. Since in our studies of 1,2,3-trichloropropane, gavage administration induced 3-fold greater concentrations of 8-OHdG, εdA, and εdC (La *et al.*, 1997).

Single Bolus Gavage Exposure To DCA

Rats and mice were exposed by gavage to determine whether bolus administration increases endogenous DNA adduct formation. Rats were exposed to a single gavage administration of DCA at 100 mg/kg, and mice to 400 mg/kg. Animals were sacrificed 4, 8, or 24 hours following exposure, and DNA was isolated from the liver for analysis.

We measured 8-OHdG by HPLC-ECD, N^2,3-εGua by GC/HRMS, and 1,N^6-εdA and 3,N^4-εdC by ^{32}P-postlabeling. We believed that theses experiments would provide information on whether endogenous DNA adducts are increased by high dose exposure to a disinfection by-product. In addition, the experiments would also indicate when maximum increases of endogenous DNA adducts occur following exposure, which would be important as we planned experiments to test other disinfection by-products for their effects on endogenous DNA lesions.

We did not observe an increase in the concentration of 8-OHdG in mouse liver DNA (Specific Aim 4). Previous studies have reported either small or no increase in 8-OHdG or lipid peroxidation from DCA exposure, depending on the dose and exposure method (Austin *et al.*, 1995, 1996; Parrish *et al.*, 1996). Larger increases have been reported for brominated acetates.

Drinking Water Exposure To DCA For 28 Days

In collaboration with Dr. Anthony DeAngelo (US EPA, Research Triangle Park, NC), F344 rats and B6C3F1 mice were exposed to DCA by drinking water for 28 days. Concentrations of DCA were 0, 0.5 or 1.5 g/L for rats, and 0 or 3.5 g/L for mice. We analyzed hepatic DNA for the reactive oxygen species (ROS) derived DNA lesion, 8-OHdG, by HPLC-ECD. We also examined DNA for the lipid peroxidation induced DNA adducts, 1,N^6-εdA and 3,N^4-εdC by immunoaffinity chromatography enriched ^{32}P-postlabeling. We did not observe an increase in these endogenous DNA lesions (Specific Aim 4). In addition, we measured AP sites in hepatic DNA of the exposed mice (Specific Aim 8). AP sites were measured by a slot-blot assay developed in our laboratory that uses a reagent that reacts with the aldehydic moiety in AP sites. There was no effect of exposure on AP site concentrations in the control group versus the exposed group.

Conclusions

High-dose exposure to chemicals can result in oxidative stress and lipid peroxidation by depleting critical cellular defenses. We have examined whether high-dose exposure to BDCM or DCA affects DNA lesions associated with oxidative stress and lipid peroxidation.

Male F344 rats and B6C3F1 mice were exposed to dichloroacetic acid via drinking water for 28 days at 0, 0.5 or 1.5 g/L for rats, and 0 or 3.5 g/L for mice. In addition, rats were exposed to a single gavage administration of DCA at 100 mg/kg, and mice to 400 mg/kg. Animals were sacrificed 4, 8, or 24 hours following exposure. We analyzed hepatic DNA for 8-OHdG, by HPLC-ECD. We also examined DNA for the lipid peroxidation induced DNA adducts, 1,N^6-εdA and 3,N^4-εdC by ^{32}P-postlabeling. We did not observe an increase in these endogenous DNA lesions. In addition, we measured AP sites in hepatic DNA of the exposed mice. We did not observe any significant difference between the control and exposed groups.

The mean values in many instances are suggestive of some cause and effect relationships, but variance in the data sets perhaps prevents the ability to make salient observations. There are large variances in the 1,N^6-εdA and 3,N^4-εdC results, which are inherent to the ^{32}P-postlabeling method and repeated measurements are unlikely to overcome this problem. Other approaches, such as mass spectrometry, which yield results with greater reproducibility are likely to reduce the variance and increase the likelihood of achieving statistical significance. We currently have a GC/MS method for the quantitation of N^2,3-εGua and 1,N^2-εGua, and a LC/MS method for the quantitation of 1,N^6-εdA. However, no additional liver tissue or DNA were available for determining adduct levels by mass spectrometry and the PAC advised us to spend more time on the other specific aims in this research plan.

ORIGINAL RESEARCH PLAN - SPECIFIC AIM 6.

Quantitate DNA Adducts Derived From Bdcm, Mutagen X, and Mucochloric Acid in Lymphocyte DNA—E. Bodes, D.K. La and J. A. Swenberg

Overview

Quantitation of DNA adducts ideally would involve the target organ; however, the target organs for many disinfection by-products are relatively inaccessible and their sampling in humans is generally not feasible. Measurement of DNA adducts for routine use as biomarkers requires tissues that can be obtained noninvasively, and usually involves an accessible surrogate tissue, such as peripheral blood. Adducts at nontarget sites must correlate quantitatively with those at target sites if they are to serve as valid biomarkers of risk.

Initially, we proposed to use lymphocyte DNA to quantify adducts derived from exposure to BDCM, MX and MCA. We planned to remove blood during sacrifice and to isolate lymphocytes for DNA adduct and AP site quantitation. However, at the group meeting on November 14, 2000 the PAC and Project Manager advised us not to proceed with this specific aim but the focus on the other specific aims in this research plan.

ORIGINAL RESEARCH PLAN - SPECIFIC AIM 7.

Quantitate Endogenous DNA Adducts (8-Hydroxydeoxyguanosine and $1,N^6$-Ethenoadenine) Excreted in Urine—E. Bodes, D.K. La and J. A. Swenberg

Overview

DNA repair mechanisms excise adducts from DNA as modified deoxyribonucleosides and bases. These repair products are water soluble, and readily excreted into the urine without further metabolism. We have developed a highly sensitive and specific method to quantitate $1N^6$-εdA in urine (Detailed Materials and Methods). We planned to isolate $1N^6$-εdA from urine by immunoaffinity chromatography using specific monoclonal antibodies developed against $1N^6$-εdA. We proposed to quantitate this adduct by HPLC-ESI/MS. Isotope dilution mass spectrometry, which employs internal standards labeled with stable isotopes to control for analyte recovery during sample preparation, is considered the most accurate method for DNA adduct quantitation. We have synthesized the $[^{15}N_5]$-εdA derivative for use as an internal standard. The detection limit obtained by HPLC-ESI/MS was 270 fmol of εdA. In the urine of control rats, εdA was detected at 21.6 ± 14.8 pmol/mL. Following administration of the vinyl chloride metabolite, chloroethylene oxide, to rats, the mean concentration of εdA in urine was 1535 pmol/ml.

We proposed to expose F344 rats to different disinfection by-products (trihalomethanes, haloacetates, MX and MCA) by single gavage administration using doses that are known to enhance formation of endogenous DNA adducts and to collect urine at 6, 12, 24, 48 and 72 hours following exposure. Based on our studies with chloroethylene oxide, excretion of $1N^6$-εdA will most likely be complete by 24 hours. In addition, we planned to expose another group of rats to lower amounts of disinfection by-products via drinking water for 1 week, during which urine would have been collected and analyzed for εdA excretion. We planned to conduct similar analysis of 8-OHdG in urine. Repair of this oxidative DNA lesion results in the excision and excretion of both 8-OHdG and 8-hydroxyguanine. Quantitation of the latter is confounded by oxidized bases consumed in the diet (Degan *et al.*, 1991). We intended to quantitate 8-OHdG in urine using a published method (Degan *et al.*, 1991) where 8-OHdG is isolated by immunoaffinity chromatography, and measured by HPLC-ECD.

We expected that immunoaffinity chromatography coupled with gas chromatography mass spectrometry or HPLC-ECD would enable sensitive detection of adducts excreted in the urine. It is not known whether excretion of adducts will increase following exposure to low concentrations of disinfection by-products in drinking water. We believed that results of animals exposed to low concentrations would provide important information whether adducts in urine can be used as biomarkers in human populations. If excretion of DNA adducts occurs following high-dose gavage administration but not after drinking water exposure to low concentrations, we would have been able to identify doses at which apparent thresholds occur. However, at the group meeting on November 14, 2000 the PAC and Project Manager advised us not to proceed with this specific aim but the focus on the other specific aims in this research plan.

ORIGINAL RESEARCH PLAN - SPECIFIC AIM 8.

Investigate the Effects of Disinfection By-Products on the Formation of Abasic Sites—E. Bodes, J. Nakamura, D.K. La and J. A. Swenberg

Overview

The excision of DNA adducts by glycosylase-mediated DNA repair or spontaneous depurination/depyrimidination results in apurinic and apyrimidinic sites (also referred to as AP or abasic sites), which are among the most prevalent lesions in DNA. The occurrence of these lesions is common, with rates of 1×10^4 and 5×10^2 per mammalian cell per day for spontaneous depurination and depyrimidination, respectively. AP sites result from cleavage of the *N*-glycosyl bond between deoxyribose and purines or pyrimidines or by spontaneous depurination (Mol *et al.*, 2000). AP sites are constantly repaired by AP endonucleases during base excision repair. However, a fraction of AP sites that escapes repair cause cell toxicity or lethality and inhibit DNA replication due to the formation of strand breaks. AP sites also induce mutations, chromosome aberrations and transcription errors. DNA glycosylases recognize a specific abnormal base and remove it from DNA leaving an AP site and an intact phosphodiester bond. The base loss may be spontaneous or can be enhanced by alkylating agents or physical treatments (i.e. irradiation) which cause base modifications, a weakening of the glycosyl bond and an increase in the release of bases (Talpaert Borle *et al.*, 1987). AP sites in the DNA template hamper RNA polymerase and cause it to pause which results in a decrease of RNA synthesis. AP sites may be by-passed by DNA polymerase resulting in misincorporation of preferentially deoxyadenosine nucleotides.

AP sites introduced by mono-functional DNA glycosylases or spontaneous depurination are processed by class II APE (AP endonucleases), which are abundant in most cells, and initiate repair by incising DNA immediately 5' to the AP sites. This cleavage is followed by a single nucleotide synthesis by β-polymerase and the remaining 5'-dRp-residue (deoxyribose phosphate residue) is removed by the dRp-lyase activity of β-polymerase and the nick is sealed by ligase. AP sites generated by bi-functional DNA glycosylases (e.g. NTH1) appear to be nicked immediately 3' to the lesions through their AP lyase activity via β-elimination (Nakamura *et al.*, 2000; Demple *et al.*, 1997).

Our studies with the carcinogen, vinyl chloride, indicate that formation of εdA (ethenodeoxyadenosine) relative to εG (ethenoguanosine) is 50-time greater, but εdA is present at 1/10th the concentration of εG *in vivo*. This suggests preferential repair of εdA by methylpurine DNA glycosylase. AP sites that result from excised εdA may account for the vinyl chloride-induced mutations at AT sites. A recent study has shown that MX primarily induces GC to TA and AT to TA transversions, consistent with mutations caused by AP sites (Hyttinen and Jansson, 1996). Hyttinen and Jansson (1995) reported that high concentrations of MX (30 mM) induce abasic sites in plasmid DNA. Nunn and Chipman (1994) proposed that MX caused the formation of etheno adducts, which are excised by methylpurine DNA glycosylase to yield AP sites. Similarly, etheno adducts induced by MCA may also produce AP sites. Many other DNA adducts removed by glycosylase-mediated DNA repair or by spontaneous depurination or depyrimidination result in AP sites. Whether DNA adducts derived from trihalomethanes yield AP sites is unknown.

Our laboratory developed a highly sensitive and specific Slot-Blot assay (Detailed Materials and Methods), which enables the detection of 2.4 AP sites/10^7 bases (Nakamura *et al.*, 1998). This assay requires a small amount of genomic DNA (15 μg) and no radioactive materials. This method involves the use of a probe that reacts with the aldehyde functional group in abasic sites. DNA is then analyzed by a ELISA-like slot-blot assay. The slot-blot assay measures the amount of heat-labile DNA adducts as well as DNA glycosylase activity.

As part of Specific Aim 8, we have investigated the effects of different disinfection by-products. We quantitated AP sites in DNA isolated from mice and rats exposed to different disinfection by-products, including BDCM (Specific Aim 1), MX (Specific Aim 2), and MCA (Specific Aim 3). In addition, we investigated whether the route of exposure to different disinfection by-products affects generation of AP sites (Specific Aims 4 and 5). We speculated that high dose exposure to disinfection by-products would increase the concentration of AP sites.

This newly developed assay is a convenient, but highly sensitive tool to detect a commonly occurring DNA lesion. Since high dose gavage administration is likely to increase the concentrations of a number of DNA adducts, including endogenous adducts, we expect increased AP sites. Whether AP sites will increase following exposure to low concentrations of disinfection by-products in drinking water is not known. Dose-response relationships will be used to identify doses at which thresholds occur.

The Effect Of Bromodichloromethane(BDCM) On AP Site Formation

Overview. BDCM and other trihalomethanes are considered to be the most prevalent by-products of chlorinated water. The trihalomethanes are metabolized by cytochrome P450 monooxygenase to reactive dihalocarbonyl intermediates that react with nucleophilic compounds and macromolecules (Melnick et al., 1998). Trihalomethanes may also undergo reductive metabolism mediated by cytochrome P450 to produce dihalomethyl radicals. Another pathway that may contribute to the carcinogenicity of trihalomethanes is where glutathione-S-transferase (GST) catalyzes the conjugation of trihalomethanes with glutathione forming DNA-reactive S-dihalomethyl glutathione metabolites (Pegram et al., 1997). Testai et al (1990) and Coffin et a. (2000) showed that the effects of the oxidative pathway intermediates were inhibited by glutathione (GSH), whereas metabolites formed by the reductive pathway were not efficiently scavenged by GSH. Thus, the induction of toxicity and carcinogenicity of the trihalomethanes may involve contributions from multiple activation pathways and different trihalomethanes may induce adverse effects via different mechanisms. It is commonly considered that chloroform is nongenotoxic in in vitro and in vivo systems, whereas BDCM is genotoxic. It has been hypothesized that the carcinogenicity of chloroform is due to its cytotoxicity effects followed by the induction of regenerative cell proliferation (Bull et al., 1995) with concurrent changes in DNA methylation (Coffin et al., 2000). The exact mechanism by which BDCM induces carcinogenesis is unknown.

Single gavage exposure study. AP sites were investigated in the liver of F344 rats exposed to BDCM. Rats were administered BDCM (0, 0.3, 1.0, or 3 mmol/kg) by corn oil gavage and sacrificed 6 hours later. We have focused on the liver in this study because of its role in metabolism, and its large yield of DNA. Keegan et al (1998) identified the acute, oral NOAEL for liver toxicity, in the case of BDCM, to be 0.5 mmol/kg. In our study, rats were exposed via single corn oil gavage to 0.3, 1.0 or 3.0 mmol/kg BDCM.

We observed no significant increase in AP sites for all exposure groups (Table 3.5). Our findings still do not explain the genotoxicity of BDCM.

Table 3.5

AP sites in hepatic DNA of F344 rats following exposure to BDCM.

Dose (mmol/kg)	AP sites (per 10^6 nucleotides)
0	11.0 ± 1.4
0.3	10.5 ± 2.1
1	11.6 ± 2.9
3	10.4 ± 2.5

Mean ± SD with $p > 0.05$

TEMPO, an antioxidant, was not added to samples during DNA extraction and the X-ray film was analyzed by an Ultrascan XL scanning densitometer.

The Effect Of Mutagen X On AP Site Formation

Overview. Most of the mutagenicity associated with chlorinated drinking water derives from MX (~60%). MX is a potent direct-acting bacterial mutagen and mammalian cell clastogen. DeMarini et al (1995) concluded that MX is one of the most potent bacterial mutagens known. Hyttinen et al (1995) found that MX induced predominantly G:C → T:A transversions in Salmonella typhimurium TA100. MX also induces both base-pair substitutions (TA100) and frameshift mutations (TA98) in bacterial DNA (DeMarini et al., 1995). Results obtained by Hyttinen et a, 1995 and Knasmuller et al (1996) in S. typhimurium TA 100 is in accordance with AP sites playing a role as intermediates in MX-induced mutagenesis since AP sites induce G:C→T:A and A:T→T:A transversions in bacteria. MX is clearly genotoxic in bacteria, but the compound appears to have relatively low genotoxic effects in mammalian cells (Holme et al., 1999; Marsteinstredet et al., 1997). The activity of MX is dependent on the existence of the α,β-unsaturated carbonyl arrangement in conjunction with the presence of the dichloromethyl group at the C4 position. An ethenoformyl adduct was formed in a reaction of MX with guanosine

($1,N^2$-ethenoformylguanosine; εfGuo) while a N^6-propenal adduct and two fluorescent cyclic $1,N^6$-propenoformyl adducts were obtained in a reaction of MX with 2'-deoxyadenosine (Munter et al., 1999 and 1998; Le Curieux et al., 1997). However, Munter et al (1999) was not able to detect the εfGuo adduct in calf thymus DNA and they speculated that the adduct may be formed at levels lower than the detection limit of their HPLC analytical method (~5 adducts /10^7). It is not known whether any these MX-induced adducts are formed in vivo, but because of its potent mutagenicity there is considerable interest in identifying DNA adducts of MX.

Single gavage exposure study. We exposed F344 rats to MX (15 mg/kg, 0.07 mmol/kg) by corn oil gavage. Animals were sacrificed 24 hours later, and liver DNA was examined for AP sites. We did not detect a statistically significant increase in AP sites (Table 3.6). Our results are in contrast with those reported by Hyttinen and Jansson (1995). These researchers found that high concentrations of MX (30 mM) induce abasic sites in plasmid (PM2) DNA. The lack of increase, found in our experiments, may be the result of: (1) the lower amount of MX administered in our study, (2) the time of exposure (24 h) may have been too short, (3) exposure to MX does not result in an increase of AP sites in animal tissues, or (4) if AP sites are induced after exposure MX their repair maybe too rapid to allow detection by our slot-blot method.

The Effect Of Mucochloric Acid On AP Site Formation

Overview. MCA appears to be present in chlorinated drinking water at concentration levels comparable to MX (3-67 ng/L; Kronberg et al., 1993). Hyttinen et al (1995) found that the MCA induced predominantly G:C→A:T transitions in Salmonella typhimurium TA100. Kronberg and colleagues (1992) demonstrated that when MCA was reacted with nucleosides under physiological conditions $1,N^2$-eGua, $1,N^6$-εA and $3,N^4$-εC adducts were produced. They suggested that the formation of the etheno adducts was due to the initial degradation of MCA to chloromalonaldehyde which then breaks down to form chloroacetaldehyde which in turn produces the etheno adducts. It has also been reported that chloroacetaldehyde adduction of DNA in vitro leads almost exclusively to mutations at cytosine residues in Escherichia coli. The mutations were predominantly G:C→A:T transitions and the adduct responsible for the mutations was $3,N^4$-εC which have properties suggestive of non-instructional DNA lesions.

Single gavage exposure study. We analyzed hepatic DNA from rats exposed to MCA. F344 rats were administered MCA (100 mg/kg, 0.6 mmol/kg) by corn oil gavage, and sacrificed 4 or 24 hours later. Additional rats were exposed to MCA (100 mg/kg, 0.6 mmol/kg) daily for 5 days, and sacrificed 4 hours after the last exposure.

Following single exposure to MCA, there was a statistically significant increase in AP sites only at the 24 hour time point, but there was no increase in AP sites following multiple exposures (Figure 3.10). The absence of an increase in AP sites following that 5 day exposure may be due to the induction of enzymes involved in the repair of AP sites. Some proteins involved in base excision repair such as APE1 (class II AP endonuclease) are induced at about 6 and 12 hours after oxidative stress in human cells at mRNA and protein levels, respectively (Ramana *et al.*, 1998).

The observed increases in AP sites are consistent with the finding by Kronberg *et al* (1992) that MCA induces the formation of etheno DNA adducts. In Specific Aim 10, we also reported that MCA produced the etheno DNA adduct, N^2,3-εGua in *in vitro* experiments. Etheno DNA adducts are substrates for *N*-methylpurine DNA glycosylase, which efficiently excises adducted bases. The increased concentrations of AP sites indicate that the resulting AP sites were not completely repaired. The lack of an increase in AP sites following multiple exposures to MCA suggests possible induction of enzymes involved in the repair of AP sites.

We have developed a battery of sensitive assays to quantify aldehydic AP sites in DNA, and we recently have developed methods to characterize AP sites induced by reactive oxygen species (Nakamura *et al.*, 2000). These assays, based on reaction of the aldehyde group in an AP site with a hydroxylamine probe bearing a biotin residue, have sensitivity to detect 1 AP site per 106 nucleotides. We have identified putrescine-excisable AP sites as major DNA lesions, which account for approximately 2/3 of total AP sites (Nakamura *et al.*, 2000). The prevalence of putrescine-excisable AP sites may be due, in part, to its direct formation by reactive oxygen species. Attack of deoxyribose by reactive oxygen species, and hydrogen abstraction are known to induce AP sites at either 5'- or 3'-termini. We will be able to distinguish 5'- and 3'-nicked AP sites from intact AP sites by incubating the DNA with T7 gene 6 exonuclease (5'-3' exonuclease) and exonuclease III (3'-5'-exonuclease) prior to AP site analysis.

Previously, we detected an increase in the number of AP sites in hepatic DNA of F344 rats following a single exposure to MCA at 100 mg/kg. Therefore, we further investigated the increase in AP sites by a newly developed AP site cleavage assay using *E. coli* exonuclease III (3'-5' exonuclease) and T7 gene 6 exonuclease (5'-3' exonuclease) on DNA extracted from rat liver harvested 24 hours after treatment of rats with MCA at 100 mg/kg.

Exonuclease III treatment reduced the original number of AP sites to 3.2 $\pm$ 1.7 lesions and 4.2 $\pm$ 2.5 lesions per 10^6 nucleotides in the control and MCA groups, respectively (Table 3.6; Figure 3.11). In contrast, T7 exonuclease excised AP sites in liver DNA were significantly different between the control and MCA-treated group. The combined fractions of intact and residual AP sites were comparable between the control and MCA groups. These data suggest that MCA increased the number of T7 Exo-excisable AP sites significantly, but not Exo III-excisable AP sites in hepatic DNA. This increase in T7 Exo-excisable AP sites may cause cell death or chromosomal aberrations.

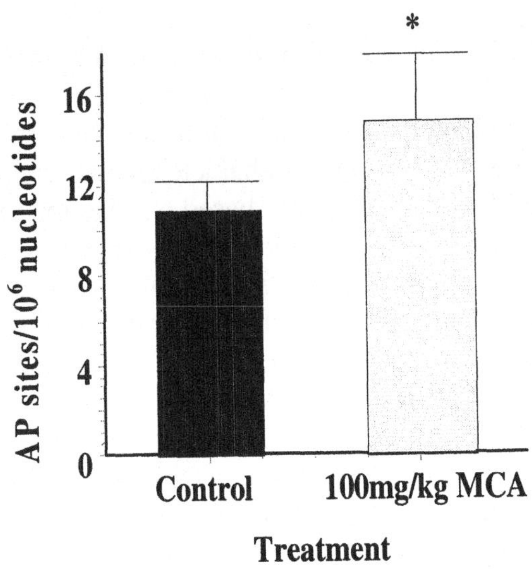

Figure 3.10. Quantitation of AP sites in F344 rat liver DNA after a single gavage dose of 100 mg/kg MCA. Rats were sacrificed 24 h afterwards. Values represent the mean ± SD with * $p < 0.05$. TEMPO was not added to the samples during DNA isolation and the X-ray film was analyzed by an Ultrascan XL scanning densitometer.

Table 3.6
The site of cleavage at AP sites in hepatic DNA of F344 rats following single exposure to 100 mg/kg MCA by corn oil gavage. Animals were sacrificed 24 hours later.

AP sites per 10^6 nucleotides	Total AP sites	T7 Exo-excisable AP sites	Exo III-excisable AP sites	Intact/Residual AP sites
Control (0 mg/kg)	10.9 ± 1.3 (n=6)	0.1 ± 0.4 (n=5)	3.2 ± 1.7 (n=5)	7.7
MCA (100 mg/kg)	14.8 ± 3* (n=6)	2.7 ± 1.9† (n=5)	4.2 ± 2.5 (n=5)	7.9

T7 exonuclease excisable AP sites; Exonuclease III excisable AP sites
TEMPO was not added to the samples during DNA isolation and the X-ray film was analyzed by an Ultrascan XL scanning densitometer. Mean ± SD with * $p < 0.05$ and †$p < 0.01$.

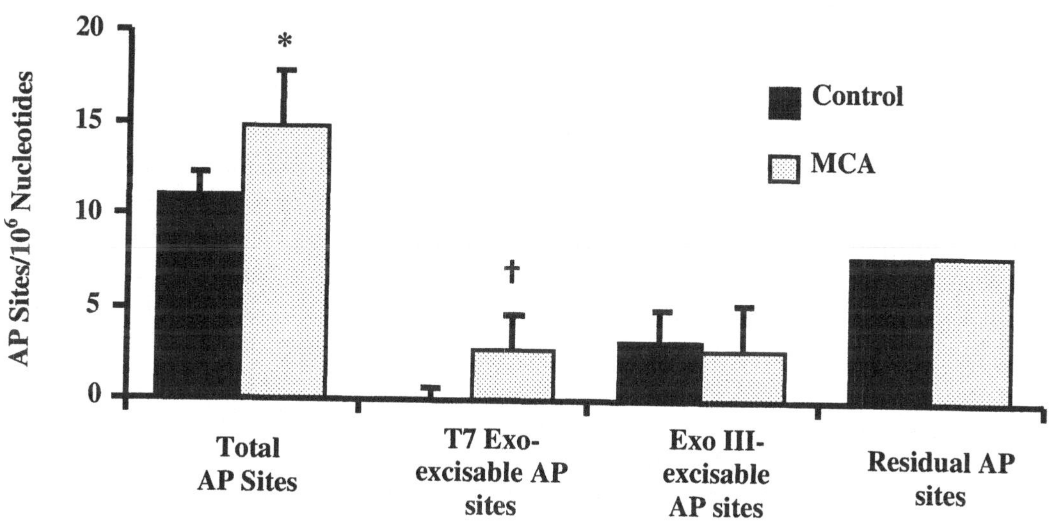

Figure 3.11. The site of cleavage at AP sites in hepatic DNA of F334 rats following single exposure to 100 mg/kg MCA by corn oil gavage. Rats were sacrificed 24 hours later. TEMPO was not added to the samples during DNA isolation and the X-ray film was analyzed by an Ultrascan XL scanning densitometer.

Data represents mean ± SD with * $p < 0.05$ and † $p < 0.01$.

The increase in the number of T7 Exo-excisable AP sites could be due to an accumulation of 5'-deoxyribophosphodiester (5'-dRp) residues introduced by an incision of AP sites by class II AP endonuclease such as APE1. Recently, Wilson and coworkers demonstrated that the excision of 5'-dRp is a rate limiting process during short-patch base excision repair pathway in cells (Sobol *et al.*, 2000). Therefore, it is possible that DNA adducts introduced by MCA were efficiently repaired by a DNA glycosylase and led to regular AP site formation. Abundantly present APE1, the next enzyme in base excision repair, efficiently excised 5' to the lesions, which introduced T7 Exo nuclease-excisable AP sites in genomic DNA. Since the excision of the 5'-dRp moiety by DNA polymerase β is thought to be the rate limiting step for shot-patch base excision repair this could explain the increases in T7 Exo-excisable AP sites. An acute dose response study was conducted to determine if a dose response existed for AP site accumulation after MCA exposure (Revised Research Plan – Specific Aim 10).

The Effect Of DCA On AP Site Formation

Overview. Haloacetates are produced in the chlorination of drinking water in the range of 10-100 μg/L. As bromide concentrations increase in the water, brominated haloacetates appear at higher concentrations than the chlorinated forms. Dibromoacetate (DBA) is also produced with the use of alternate disinfectants, especially ozone. When administering brominated haloacetates to mice in drinking water (0.2-3.0 g/L) for 12 weeks, DBA was the only brominated haloacetate that consistently increased acyl CoA oxidase activity as well as cell replication in the liver. Brominated haloacetates induced prolonged oxidative stress in the livers of B6C3F1 mice and acute exposures to the brominated derivatives showed that they were more potent inducers of oxidative damage to DNA (increase in 8-OHdG), and lipid peroxidation (measured by thiobarbituric acid-reactive substances; TBARS) than the chlorinated forms (Parrish et al., 1996; Austin et al., 1995 and 1996; Larson and Bull, 1992). The bromine-carbon bond is more labile than the chlorine-carbon bond which may explain why brominated haloacetates showed a greater increase in lipid peroxidation and 8-OHdG formation (Austin et al., 1996; Parrish et al., 1996). Since it is already known that DBA induces oxidative stress we decided to focus our efforts on DCA in stead. In addition, liver tissue form mice exposed to DCA became available to our research group and thus our goal was to determine if AP site formation was induced after exposure to DCA in drinking water.

Chronic drinking water study. We measured AP sites in hepatic DNA of B6C3F1 mice exposed to 0 or 3.5 g/L DCA in drinking water for 28 days. AP sites were measured by a slot-blot assay. There was no effect of exposure on AP site formation in the control group versus the exposed group (Table 3.7).

Table 3.7
AP sites in hepatic DNA of B6C3F1 mice following a 28 day exposure to DCA in drinking water.

	AP Sites (per 10^6 nucleotides)
Control	8.3 ± 2.8
Exposed	7.4 ± 4.0

Study conducted in collaboration with Dr. DeAngelo at the US EPA. TEMPO was added to the samples during DNA isolation and the X-ray film was analyzed by a new Kodak Image analysis system.
Mean ± SD with $p > 0.05$.

Conclusions

AP sites result from cleavage of the *N*-glycosyl bond between deoxyribose and purines or pyrimidines or by spontaneous depurination. AP sites are constantly repaired during base excision repair and fractions that escape repair can cause cell toxicity or lethality, inhibit DNA replication, induce mutations, chromosome aberrations and transcription errors.

We have developed a highly sensitive and specific slot-blot assay which uses a probe that reacts with the aldehyde functional group in abasic sites. The slot-blot assay measures that amount of heat-labile DNA adducts as well as DNA glycosylase activity. We investigated the effects of different disinfection by-products on the formation of AP sites.

We found no significant difference in AP site formation between the control and BDCM treated groups. F344 rats received a single gavage dose of BDCM (0, 0.3, 1.0 or 3.0 mmol/kg) mixed with corn oil and were sacrificed 6 hours later. Our findings are consistent with current literature that does not support an induction of any DNA adducts or AP sites after exposure to BDCM. Thus, we were not able to explain the genotoxicity of BDCM.

We observed no increase in AP site formation in liver DNA between control and either the F344 rats or Eker rats after exposure to MX (Specific Aim 9). The F344 rats received a single dose (15 mg/kg; 0.07 mmol/kg) by corn oil gavage and were sacrificed 24 hours later. The Eker rats were exposed via drinking water to MX at 0.007 g/L and 0.07 g/L for 4 months. In contrast to our results, Hyttinen and Jansson (1995) reported that abasic sites were inducted in plasmid DNA with high concentrations of MX (30 mM). The lack of an increase, as in our experiments, may be due to the lower amount of MX administered in our study. However, the doses we employed are orders of magnitude higher than that found in drinking water. Thus, the relevance of the Hyttinen and Jansson (1995) study for human exposure is questionable.

We observed a significant increase in AP site formation in rat liver DNA after a single gavage dose of MCA (100 mg/kg). In this study, MCA was dissolved in DMSO and mixed with corn oil, and the rats were sacrificed 24 hours after receiving the single dose. By using the AP site cleavage assay, developed in our laboratory by Jun Nakamura, we tried to determine the site of cleavage. We found a significant increase in the formation of T7 Exo nuclease-excisable AP sites between the control and treated groups. The AP sites were primarily cleaved 5' to the AP site, suggesting they are not due to reactive oxygen species. The increase may be due to an accumulation of 5'-dRp residues introduced by an incision of AP sites by class II AP endonucleases. How MCA interferes with the function of base excision repair enzymes is not known, but it may affect one of the rate limiting steps for shot-patch base excision repair. In a follow-up study we proposed to determine if a dose response existed for AP site formation after receiving a single gavage dose of 0, 10, 30, 100 or 300 mg/kg (Specific Aim 10). We decided to mimic human exposure by dissolving MCA in distilled deionized water. Unexpectedly after 24 hours all the rats in the 300 mg/kg and one from the 100 mg/kg dose groups were found dead. In addition there was no statistically significant difference in the amount of AP site formation between the control and MCA exposed animals. We are currently planning a larger study to determine if the type of gavage vehicle plays a role when animals are exposed to MCA. We plan to conduct an extensive necropsy where we will collect tissue samples for histology, serum chemistry and samples for the measurement of DNA adducts and other lesions to better understand why these animals died. Our objective is to focus initially on the liver and kidney as well as an extensive clinical chemistry panel.

Haloacetates induce oxidative DNA damage and lipid peroxidation when administered by gavage (Larson and Bull, 1992; Austin *et al.,* 1995, 1996). In addition, brominated haloacetates induced prolonged oxidative stress in the livers of B6C3F1 mice after acute exposures, whereas DCA and TCA failed to produce an increase (Parrish *et al.*, 1996). Our goal was to determine if AP site formation was induced after exposure to DCA in drinking water for 28 days. We measured AP sites in liver DNA from B6C3F1 mice exposed to 0 or 3.5 g/L and we observed no significant difference between the control and treated group. Thus our results are consistent with current literature in that DCA exposure via drinking water may not play a significant role in the induction of oxidative stress.

REVISED RESEARCH PLAN – SPECIFIC AIM 9.

Identify and Quantitate Direct and Indirect DNA Adducts and Lesions Induced by Mutagen X in Eker Rat Liver DNA—E. Bodes, K. Jayaraj, H. Koc, A. Gold and J. A. Swenberg

Overview

We proposed to isolate and quantitate a MX specific ethenoformyl adduct. In addition, we proposed to investigate other DNA adducts and lesions (i.e. 8-OHdG, AP sites and etheno adducts) in hepatic DNA from Eker rats after chronic exposure to MX via drinking water for 4 months. We expected that the HPLC-ESI/MS/MS method would be a very sensitive and specific tool to enable detection of the ethenoformyldeoxyguanosine DNA adduct derived from MX. We also believed that the IA-GC/HRMS method would be a powerful tool for the analysis of MX induced N^2,3-εGua and 1,N^2-εGua adducts due to extreme sensitivity and specificity. We speculated that high-dose chronic drinking water exposure would enhance concentrations of endogenous DNA adducts.

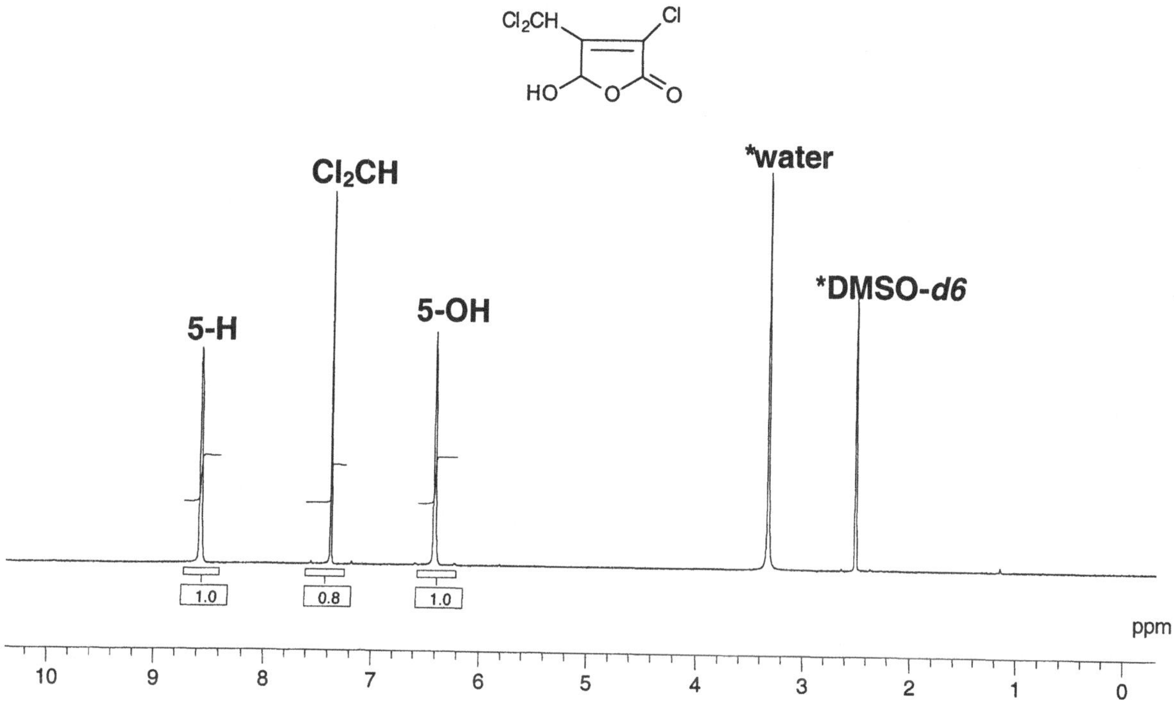

Figure 3.12. ^{1}H-NMR spectral data of MX.

Synthesis And Quantitation Of Ethenoformyldeoxyguanosine Adduct Derived From Mutagen X

We decided to perform the synthesis of εfdGuo on large scale since MX became readily available in large quantities to our research group through collaboration with Dr. Wolf (US EPA, Research Triangle Park, N.C.). We proposed to purify εfdGuo by a combination of analytical and preparative reverse phase chromatography, and to characterize the adduct by UV spectral analysis, proton NMR and ESI/MS/MS. The purity of MX (7 mg dissolved in 0.75 mL solvent) and 2'-deoxyguanosine (2'-dGuo; 3.5 mg dissolved in 0.75 mL solvent) were checked by ^{1}H-NMR (nuclear magnetic resonance) in DMSO-d_6. The purity of 2'-dGuo and MX was at least 98% as estimated by ^{1}H-NMR analysis (Figures 3.12 and 3.13).

MX (358 mg, 1.65 mmol) was reacted with 2'-dGuo (220 mg, 0.82 mmol) in 145 mL of 0.01M phosphate buffer solution (pH 8.2). The pH of the reaction was checked daily and readjusted with 1M NaOH. The reaction was performed at 37°C for 7 days and at completion the reaction mixture was cooled overnight at 4°C. The mixture was filtered, dried via lyophilization and the resulting light brown solid was dissolved in distilled deionized water. We used analytical HPLC to separate the εfdGuo adduct from the unreacted starting materials and other impurities (Detailed Materials and Methods). The fraction eluting at 15.6 minutes had the same UV-visual spectrum of the adduct reported by Munter *et al* (1999; Figure 3.14). We conducted analytical and preparative HPLC on a liquid chromatographic system with an UV diode-array detector. The UV spectra of the compounds were recorded with the diode-array detector, set at 360 nm, as the peaks eluted from the HPLC column. The compounds were isolated and collected from the reaction mixture by column chromatography (Detailed Materials and Methods). The mixture was desalted using the preparative C18 column, fractions of interest were collected and evaporated to dryness via lyophilization. The residue was dissolved in distilled deionized water and was subjected to spectroscopic and spectrometric studies (Figure 3.15). ^{1}H-NMR spectrum of the adduct in DMSO-*d6* is shown in Figure 3.16. The ^{1}H-NMR signal assignments were based on H-H and C-H correlational data and chemical shifts. The solute was dissolved in distilled deionized water for liquid chromatography with electrospray ionization tandem mass spectrometry (Detailed Materials and Methods).

^{1}H-NMR spectroscopic, HPLC-diode array and HPLC/MS data. The ^{1}H-NMR spectral data for MX were as follows (Figure 3.12): δ 8.6 (s, 1H, H-5), 7.4 (s, 1H, -Cl_2CH) and 6.4 (s, 1H, HO-5). The ^{1}H-NMR spectral data for 2'-dGuo were as follows (Figure 3.13): δ 10.6 (s, 1H, NH-1), 7.9 (s, 1H, H-8), 6.45 (d, 2H, -NH_2), 6.1 (s, 1H, H-1'), 5.25 (s, 1H, OH-5'), 4.9 (s, 1H, HO-3'), 4.35 (s, 1H, H-3'), 3.8 (s, 1H, H-4'), 3.5 (m, 2H, H-5', H-5") and 2.2 (s, 1H, H-2").

The isolated εfdGuo adduct had the following spectral characteristics: UV spectrum (Figure 3.14) with a UV_{max} 230 and 360 nm, UV_{min} 312 nm with a shoulder between 240 and 270 nm. The positive ion electrospray mass spectrum (Figure 3.15) showed the following ions *m/z* (relative abundance, formation): 204 (41 %; MH^+ - deoxyribosyl + H), 290 (44 %; MH^+ - CHO +H) and 319.7 [63 %; MH^+]. The ^{1}H-NMR spectral data for the εfdGuo adduct were as follows (Figure 3.16; Table 3.8): δ 10.6 (d, 1H, CHO), 8.0 (s, 1H, H-6), 7.9 (s, 1H, H-2), 6.2 (m, 1H, H-1'), 5.4 (dd, 1H, HO-5'), 5.3 (d, 1H, HO-3'), 4.4 (m, 1H, H-3'), 3.8 (bs, 1H, H-4'), 3.5 (m, 1H, H-5'), 3.28 (s, 1H, H-5"), 3.0 (ddd, 1H, H-2') and 2.2 (ddd, 1H, H-2").

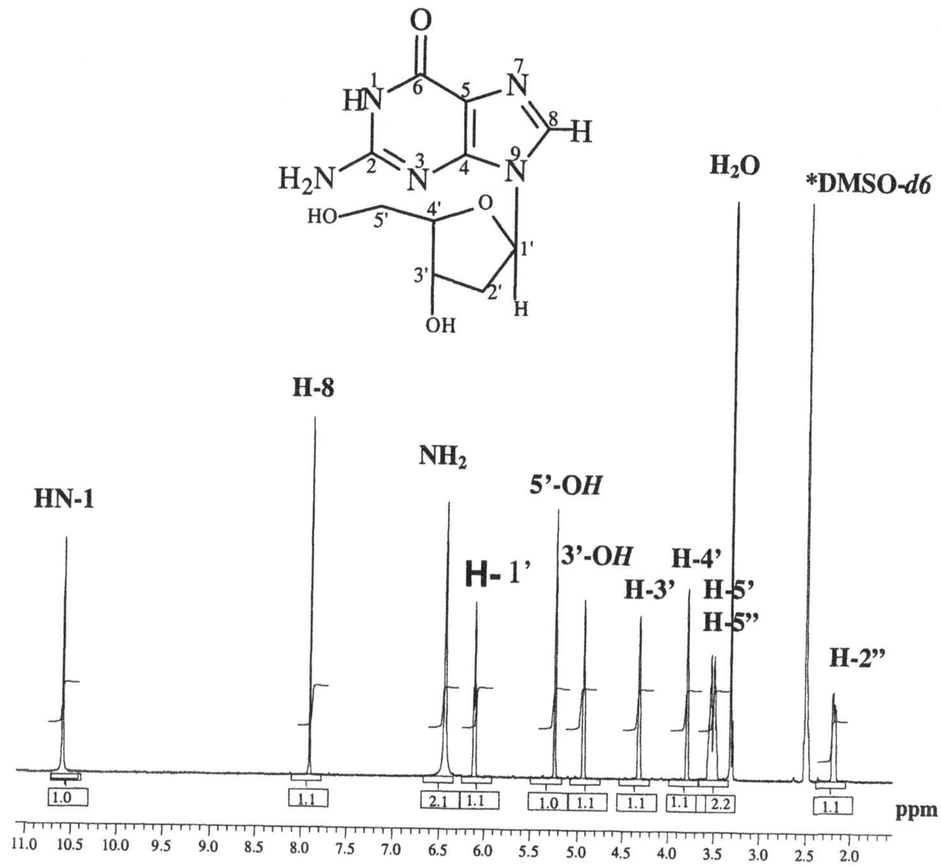

Figure 3.13. ^{1}H-NMR spectral data of 2'-dGuo.

A product of the reaction mixture was isolated by preparative chromatography and its structure was determined by UV absorbance, ^{1}H-NMR spectroscopy and mass spectrometry. The spectral data indicated that the isolated adduct was impure. An additional reaction was performed at 50°C to improve the yield. HPLC analysis indicated that the yield of the product has improved from 2.75% to 4.15% calculated by the area of the 15.7 minute peak for the adduct. Our aim was to determine if such an adduct is formed in calf thymus DNA incubated with MX and in liver samples of rats exposed to MX via drinking water. However, we were not able to develop a method that could yield large amounts of relatively pure adduct for the synthesis of analyte standards and isotopically labeled internal standards to be used in quantitative studies. Subsequently, we were unable to use liquid chromatography with electrospray ionization tandem mass spectrometry to investigate the presence of εfdGuo adducts in calf thymus DNA exposed *in vitro* to MX and DNA from male Eker rats exposed to MX.

Table 3.8
^{1}H Chemical Shifts (δ) of Protons in εfdGuo

Proton	δ (relative to TMS)	Multiplicity
H-2 (1H)	7.9	Singlet (s)
H-6 (1H)	8.0	Singlet (s)
C*H*O (1H)	10.6	Doublet (d)
H-1' (1H)	6.2	Multiple (m)
H-2' (1H)	3.0	Doublet-doublet-doublet (ddd)
H-2" (1H)	2.2	Doublet-doublet-doublet (ddd)
H-3 (1H)	4.4	Multiple (m)
H-4' (1H)	3.8	Broad-singlet (bs)
H-5' (1H)	3.5	Multiple (m)
H-5" (1H)	3.3	Singlet (s)
5'-O*H* (1H)	5.4	Doublet-doublet (dd)
3'-O*H* (1H)	5.3	Doublet (d)

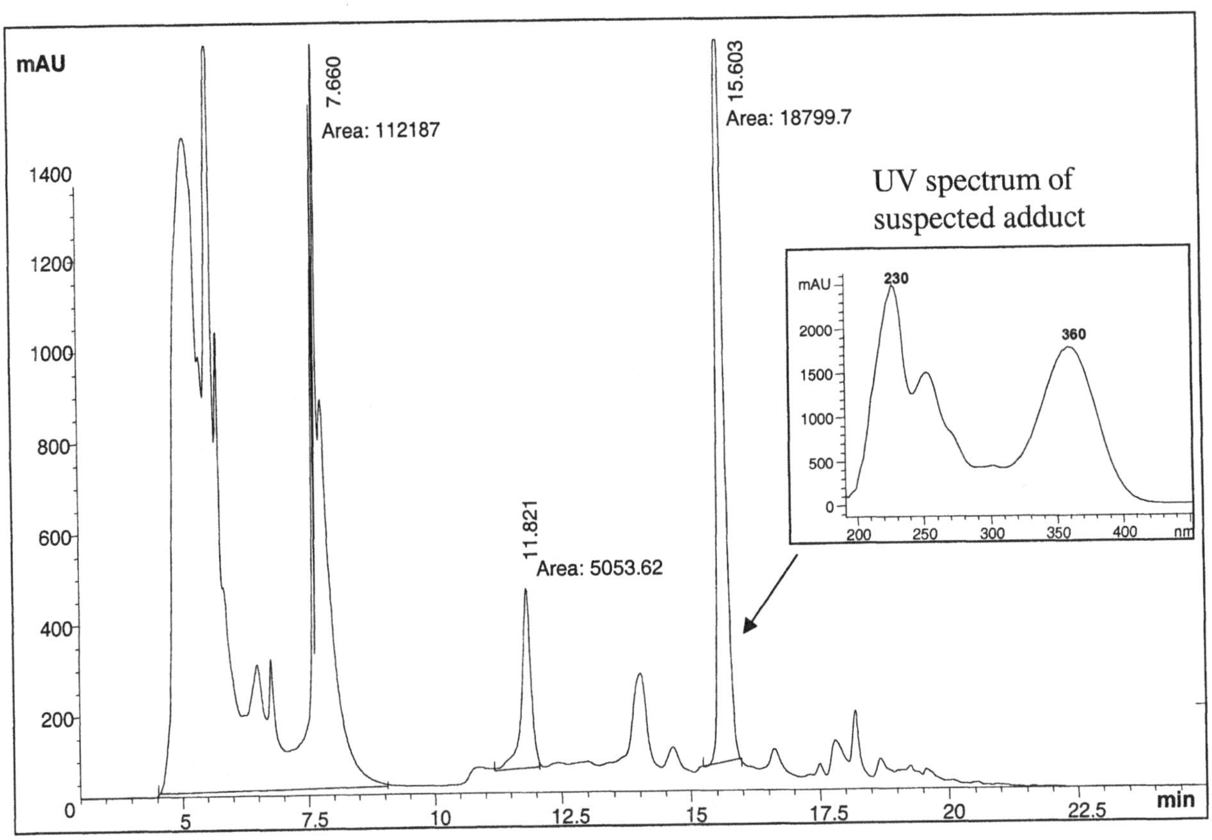

Figure 3.14. HPLC separation of the MX-2'dGuo reaction mixture.

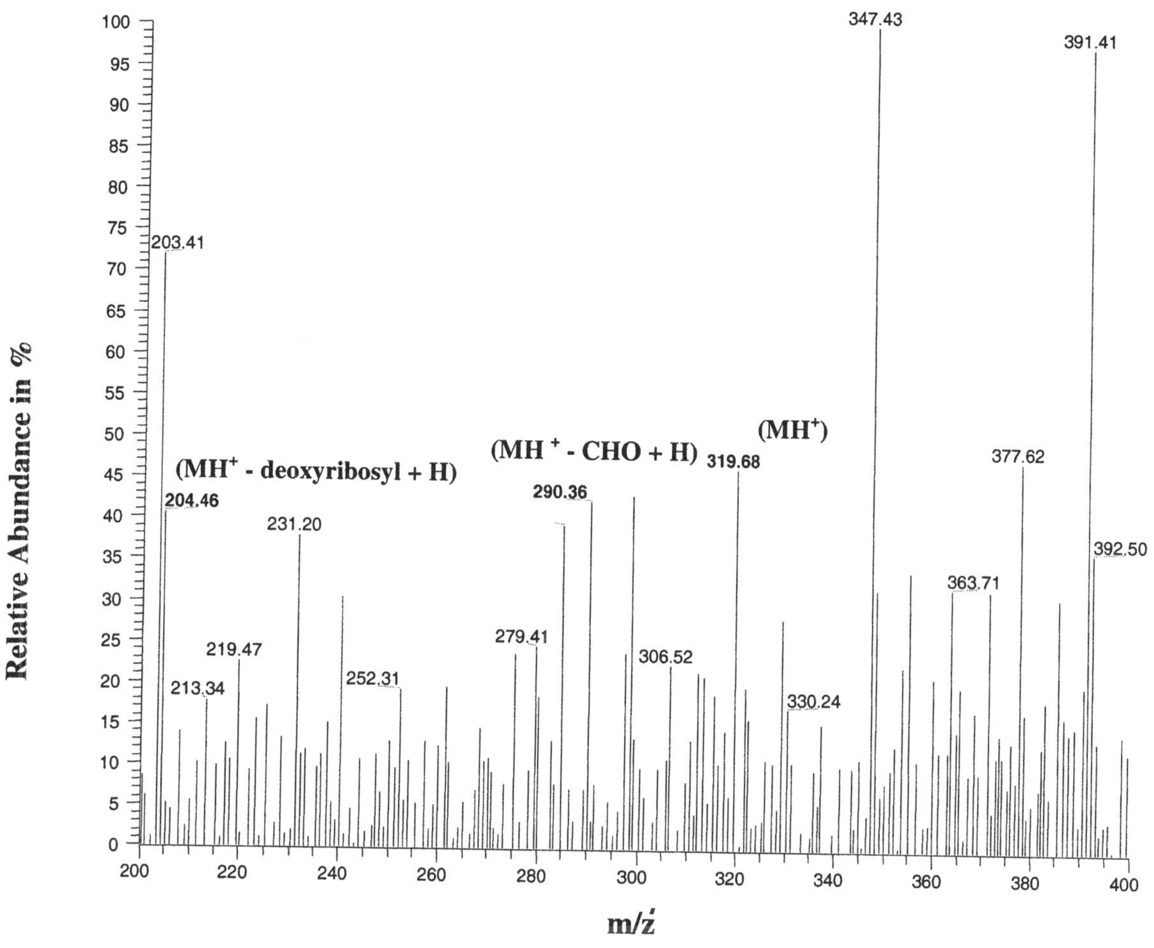

Figure 3.15. ESI/MS spectrum of the 15.6 min fraction collected by HPLC (see Figure 3.14).

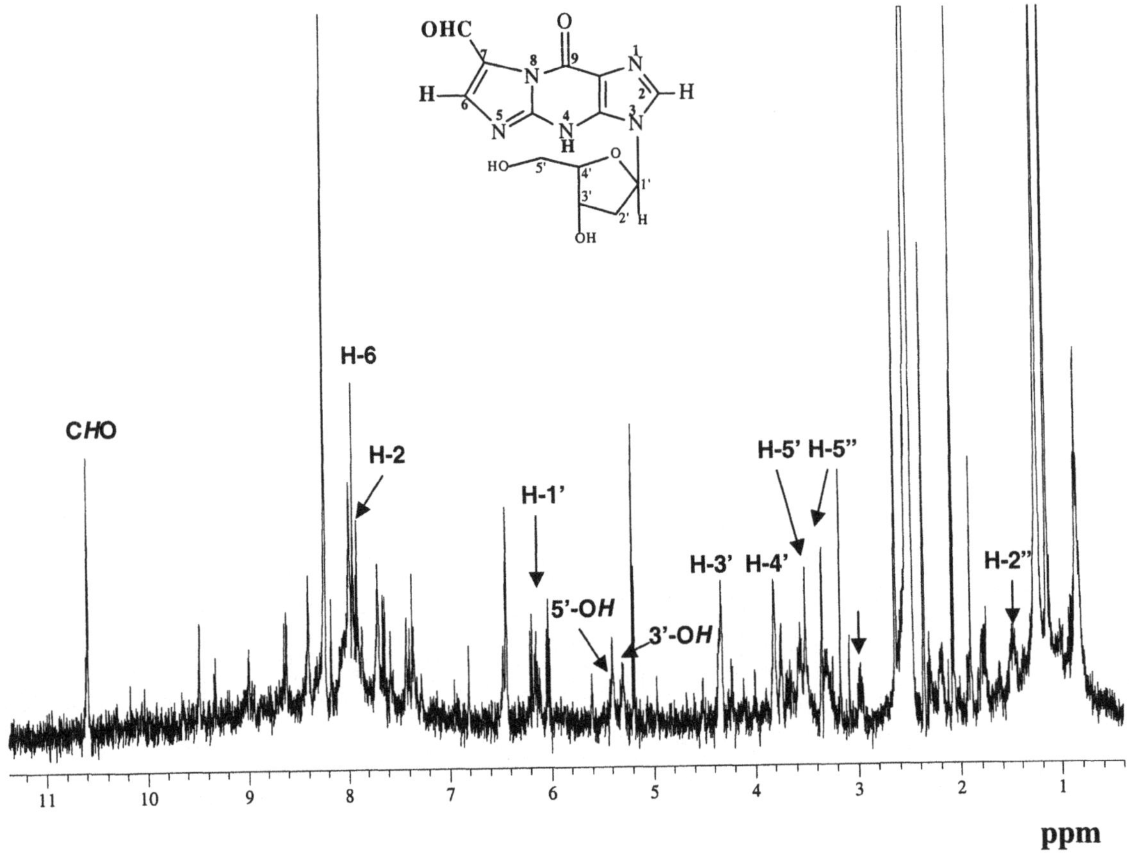

Figure 3.16. ^{1}H-NMR spectral data of εfdGuo from a reaction mixture of MX and 2'-dGuo.

Analysis Of Tissues From Chronic Drinking Water Study

Overview. An extensive and unique set of tissue samples became available for this project through a collaborative effort between our laboratory and Dr. Douglas Wolf with the National Health and Environmental Effects Research Laboratory at the US EPA (US EPA/NHEERL, Research Triangle Park, N.C.). This study was designed to examine the enhancing effect of individual and a mixture of drinking water disinfection by-products in a rat model of hereditary renal cell carcinoma (the Eker rat). Because this animal model is well defined in regard to both lesion development and genetic mutation, it allows for direct quantitation of carcinogenic response and production of statistically powerful data to understand the relative toxicity of chemicals and genetic alterations that may be chemical specific. Eker rats were exposed via drinking water to low and high doses of individual or a mixture of renal carcinogens and/or nephrotoxicants for 4 months. We were particularly interested in the Eker rats that were exposed via drinking water to MX, at a low concentration 0.007 g/L and a high concentration of 0.07 g/L. We planned to focus our attention initially on the liver for quantitation of 8-oxoguanine, apurinic/apyrimidinic (AP) sites and the ethenoguanine (εGua) adducts.

Quantitation of 8-oxoguanine in liver DNA by HPLC-ECD. We are currently using a more direct approach to the identification of the oxidized DNA base 8-oxoguanine. The original method measured the nucleoside of the adduct following non-selective enzymatic digestion of DNA. The new assay is based on a protocol published by Beckman et al (2000). This protocol involves the direct analysis of the oxidized base following selective enzymatic digestion of DNA using the base excision repair enzyme formamidopyrimidine glycosylase (Fpg). We have modified this protocol and validated the assay in our laboratory.

Artifactual generation of 8-oxoguanine is minimized during DNA isolation by the addition of radical scavengers (e.g. 2,2,6,6-tetramethylpiperidinoxyl; ICN Biomedicals Inc., Aurora, Ohio). DNA is resuspended in Fpg buffer, Fpg enzyme and distilled deionized water. After incubation and centrifugation the filtrate is injected onto a HPLC column and analyzed for 8-oxoguanine by electrochemical detection (ESA Inc., Chelmsford, Mass.). HCl is added to the retentate and after incubation the hydrolysate is analyzed by HPLC-UV for the quantitation of guanine. Data from 8-oxoguanine and guanine analysis is used to generate the number of 8-oxoguanine lesions per 10^6 guanine (Detailed Materials and Methods).

The 8-oxoguanine peak is detected and integrated at the 250 mV channel. The reaction mixtures are chromatographed on a 5 μm, 4 x 250 mm reversed phase C18 Beckman Ultrasphere ODS column (Beckman Coulter Inc., Fullerton, CA). The column is eluted isocratically with 50 mM potassium phosphate buffer, pH 5.5 with 2.0% HPLC-grade methanol at a flow rate of 1 mL/min.

Liver DNA samples from control (n=5) and treated (n=5; 0.07 g/L MX) Eker rats were mixed with Fpg enzyme and Fpg buffer, and filtered through Microcon filtration tubes. The filtrate was either stored at -80°C until analysis or injected onto the HPLC column for analysis. Similarly after exposing the retentate to acid, the hydrolysate was loaded into HPLC injection vials and stored at -80°C for further analysis by HPLC-UV for guanine quantitation.

No detectable amounts of 8-oxoguanine were observed in liver DNA from the control and the exposed groups analyzed (Table 3.9). The level of 8-oxoguanine in these animals was below our detection limit of 5 8-oxoG/10^7G when 25 μg of DNA was applied onto the column.

Table 3.9
Quantitation of 8-oxoguanine by HPLC-ECD in rat liver DNA. Eker rats were exposed to MX for 4 months via drinking water.

Treatment	Control	0.07 g/L MX
8oxoG/10^6 G	ND	ND

ND = non-detection
Detection Limit: 5 8-oxoG/10^7 G with 25 μg DNA applied onto the column.

Quantitation of AP sites in liver DNA by slot-blot assay. AP sites are the most common form of endogenous DNA damage and are constantly repaired. However, fractions that escape repair cause cell toxicity or lethality and inhibit DNA replication due to the formation of strand breaks. A sensitive slot-blot assay, developed in our laboratory, was used to quantitate abasic sites in liver DNA (Nakamura et al., 1998) from the Eker rats (n=5) chronically exposed via drinking water to MX (0.07 g/L) for four months, and from control male rats (n=5) from the US EPA study.

There was no statistically significant difference, in the amount of AP sites, between the control rats and the rats exposed to MX (0.07g/L) in drinking water for 4 months (Figure 3.17.). These results are in accordance with a previous study where F344 rats received a single exposure of MX (15 mg/kg; 0.07 mmol/kg) by corn oil gavage. The animals in this gavage study were sacrificed after 24 hours (Specific Aim 8).

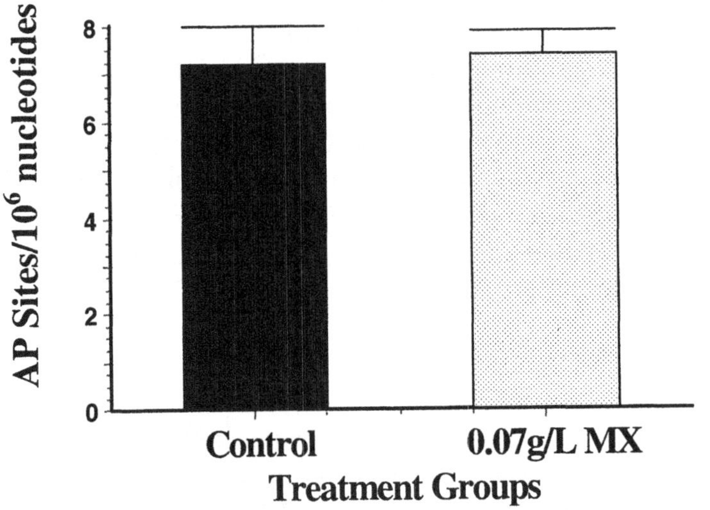

Figure 3.17. Quantitation of AP sites in liver DNA from Eker rats exposed to 0.07g/L MX in drinking water for 4 months. Data represents mean ± SD with $p > 0.05$. TEMPO was added to samples during DNA extraction and the X-ray film was analyzed by a new Kodak Image Analysis system.

Quantitation of etheno adducts in liver DNA by IA-GC/HRMS. A sensitive and selective assay was developed in our laboratory (Ham et al., 1999) for the quantitation of N^2,3-ethenoguanine using immunoaffinity (IA) chromatography coupled with gas chromatography and high resolution mass spectrometry (GC/HRMS). This method was adapted in Morinello et al (2001) to be used for the simultaneous quantitation of N^2,3-ethenoguanine (N^2,3-εGua) and 1,N^2-ethenoguanine (1,N^2-εGua). Concentrations of the promutagenic ethenoguanine DNA adducts were measured in in vitro reactions of calf thymus DNA with MX, as well as in liver DNA from control and MX-exposed male Eker rats (n = 5 per group). The Eker rats were exposed via drinking water to MX (0.07 g/L) for four months.

We reacted calf thymus DNA (Sigma; 200 μg) with 2.3 μmol MX in a total volume of 1.5 mL for 45 min at 37 °C and analyzed aliquots of this sample corresponding to 3, 10, 50, and 100 μg DNA for N^2,3-εGua and 1,N^2-εGua adducts (Detailed Materials and Methods). Following DNA hydrolysis we purified the hydrolysate by Centricon 10 filtration and we used preconditioned immunoaffinity columns for adduct purification. Following a silica gel solid phase extraction step the adduct was quantified by GC-HRMS. For adduct quantitation we used a HP5890 gas chromatograph coupled to a Varian VG250SEQ II high-resolution mass spectrometer (HRMS) operated in negative ion-chemical ionization mode. We used selected ion monitoring at m/z = 334.0351 and 337.0453 to monitor the $[M-201]^-$ fragments of 3,5-PFB_2-1,N^2-εGua and 3,5-PFB_2-$[^{13}C_3]$1,N^2-εGua, respectively. In addition, the $[M-181]^-$ fragments of 3,5-PFB_2-N^2,3-εGua and 3,5-PFB_2-$[^{13}C_4,^{15}N_2]$$N^2$,3-εGua were monitored at m/z = 354.0414 and 360.0489, respectively. We quantitated each adduct by comparing its peak area with that of the corresponding internal standard and we normalized adduct concentrations to the amount of guanine measured in each sample by HPLC.

Neither adduct appeared to be produced from the *in vitro* reaction of MX with calf thymus DNA (Figure 3.18). The small amount of N^2,3-εGua present was attributable to the endogenous background presence of this adduct in DNA. Consistent with the *in vitro* results, neither adduct were induced in rat liver DNA following exposure to MX in drinking water (0.07 g/L) for 4 months (Figure 3.19). We reanalyzed selected samples by using the new methodology for 1,N^2-εGua, but this adduct was not detected.

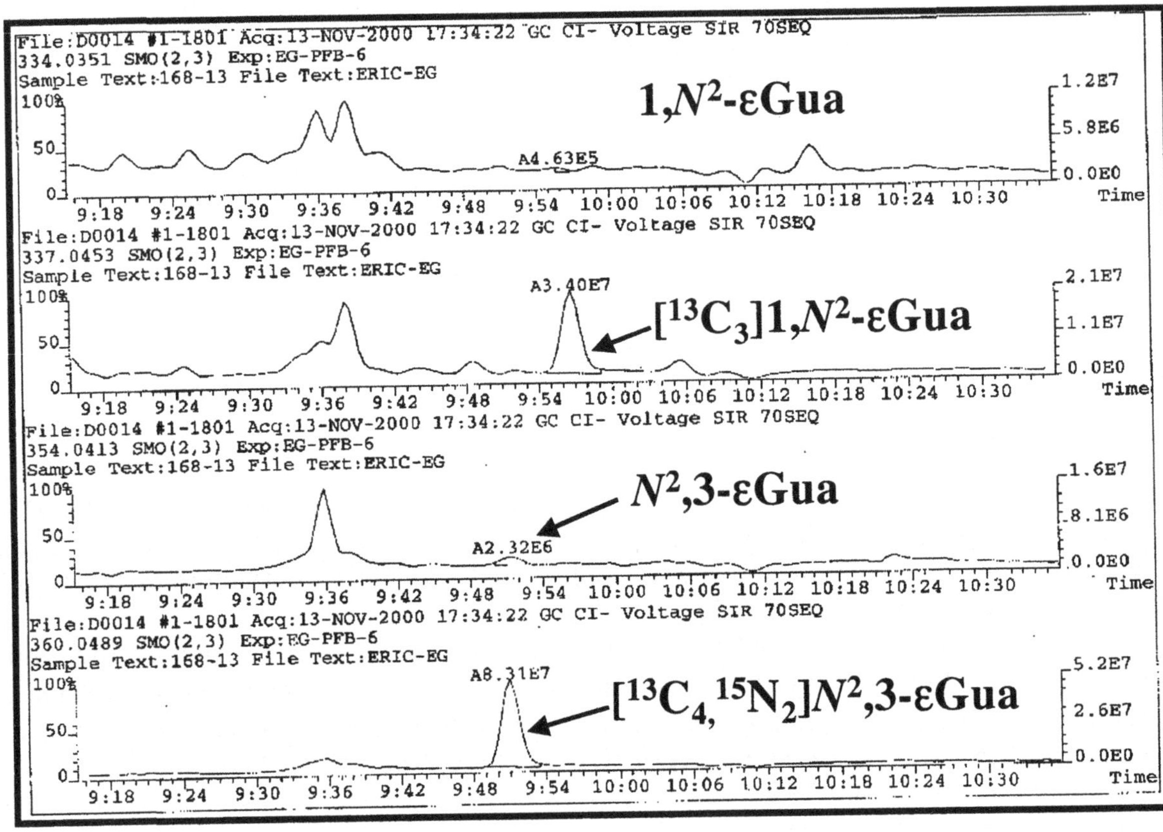

Figure 3.18. 1,N^2-εGua and N^2,3-εGua in 100 μg calf thymus DNA treated with MX *in vitro*. No 1,N^2-εGua (top) or N^2,3-εGua (middle-bottom) was induced following this exposure. The small amount of N^2,3-εGua present is due to the endogenous presence of this adduct in calf thymus DNA. The corresponding internal standards are also shown.

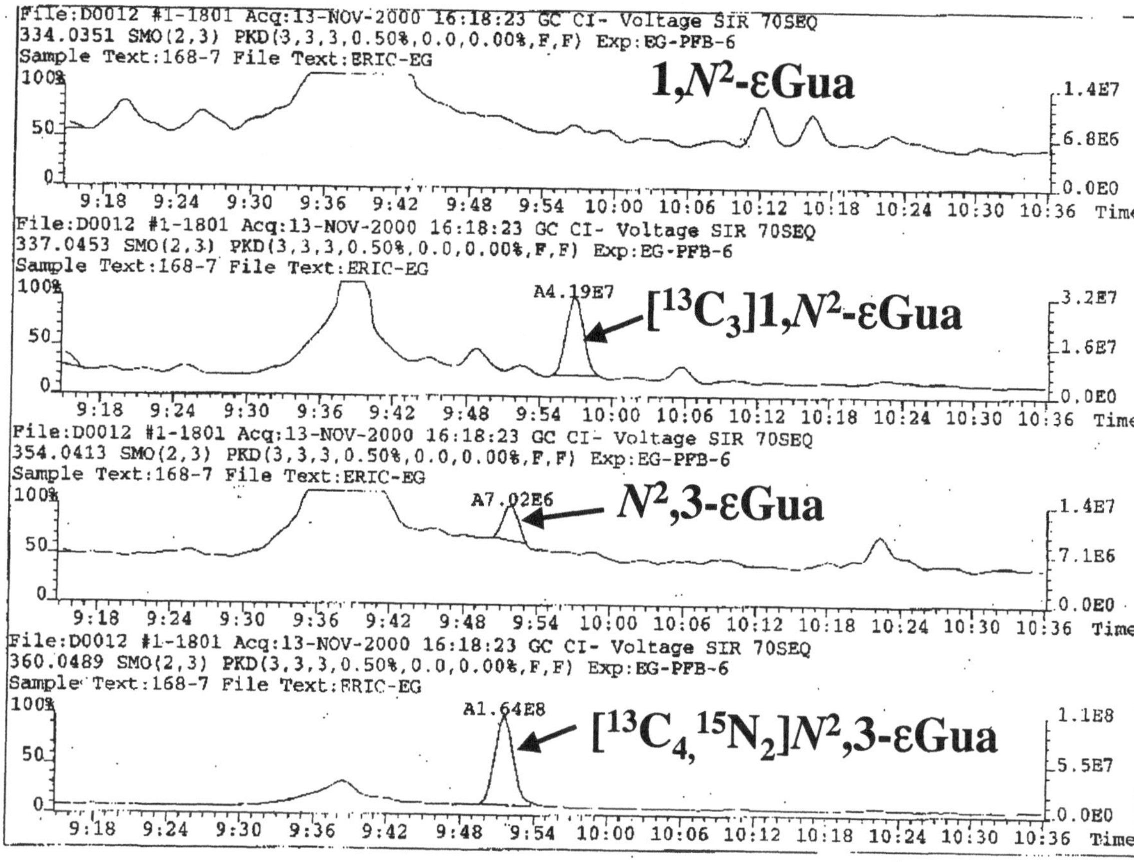

Figure 3.19. 1,N^2-εGua and N^2,3-εGua in DNA from a rat exposed to MX for 4 months. No 1,N^2-εGua (top) or N^2,3-εGua (middle-bottom) was induced following this exposure. The small amount of N^2,3-εGua present is due to the endogenous presence of this adduct in rat DNA. The corresponding internal standards are also shown.

Conclusions

After the initial promising results (see Original Research Plan - Specific Aim 2) we decided to perform the synthesis of εfdGuo on a large scale to be used in our future experiments. Our aim was to determine if such an adduct is formed in calf thymus DNA incubated with MX and in liver samples of rats exposed to MX via drinking water. Additional reactions were performed at 50°C to improve the yield. HPLC analysis indicated that the yield of the product improved from 2.75% to 4.15% calculated by the area of the 15.7 minute peak for the adduct but the spectral data indicated that the isolated adduct was impure. After several months of extensive research we were not able to develop a method that could yield large amounts of relatively pure adduct to be used in quantitative studies for the synthesis of analyte standards and isotopically labeled internal standards. Subsequently, we were unable to use a quantitative LC/ESI/MS/MS method to investigate the presence of εfdGuo adducts in calf thymus DNA exposed *in vitro* to MX and DNA from male Eker rats exposed to MX.

An extensive tissue sample set became available for this project through a collaborative effort with Dr. Wolf (US EPA/NHEERL, Research Triangle Park, N.C.). Analysis of this tissue set was outside the scope of the original proposal. In this study Eker rats were exposed via drinking water to low and high doses of individual or a mixture of renal carcinogens and/or nephrotoxicants. We were particularly interested in the Eker rats exposed to MX. We focused our attention on liver samples for 8-oxoguanine, AP sites and ethenoguanine adducts. We quantitated 8-oxoguanine by HPLC-ECD following selective enzymatic digestion of DNA by the base excision repair enzyme formamidopyrimidine glycosylase (Fpg). No detectable amounts of 8-oxoguanine were observed in liver DNA from control and exposed groups analyzed. We used the sensitive slot-blot assay, developed by Jun Nakamura, to quantitate aldehydic DNA lesions in liver DNA from F344 rats (single gavage; 0 or 15 mg/kg) and Eker rats (drinking water; 4 months; 0 or 0.07 g/L). We found no significant difference between the control and exposed rats in both studies. In addition, neither N^2,3-εGua nor 1,N^2-εGua adducts were found in the *in vitro* reaction of MX with calf thymus DNA or in liver DNA following exposure to MX in drinking water for 4 months. These combined results suggest that exposure to MX at doses used in these studies may not be high enough to induce oxidative stress.

REVISED RESEARCH PLAN – SPECIFIC AIM 10.

Identify and Quantitate DNA Adducts and Lesions Induced by Mucochloric Acid in F344 Rat Liver DNA—E. Bodes, J. Nakamura, B. F. Pachkowski, E. J. Morinello and J. A. Swenberg

Overview

We proposed to investigate several specific affects of MCA on F344 liver DNA in an acute dose response study. We planned to use a very sensitive and specific tool to enable detection of the etheno guanine DNA adducts. We also believed that the IA-GC/HRMS method would be a powerful tool for the analysis of MCA induced N^2,3-εGua and 1,N^2-εGua adducts due to extreme sensitivity and specificity. In addition, we speculated that the dose response study would enable us to detect doses at which thresholds occur after MCA exposure.

Quantitate Ethenodeoxyguanosine Adducts

The concentrations of the promutagenic ethenoguanine DNA adducts were measured in *in vitro* reactions of deoxyguanosine with MCA. The formation of 1,N^2-εGua from the reaction of MCA with deoxyguanosine (Kronberg *et al., 1992)* had been reported. The method of Ham *et al* (1999) was initially used to measure N^2,3-εGua in these samples. We also used the expanded assay that allows for the simultaneous quantitation of 1,N^2-εGua and N^2,3-εGua (Morinello *et al.*, 2001) to reanalyze some of these samples.

MCA (3.75 mmol) was reacted with deoxyguanosine (1.9 mmol) in 50 mM potassium dihydrogen phosphate buffer for 48 hr at 50 °C. The sample was diluted 100-fold and hydrolyzed to the corresponding nucleobases by incubation in 0.2 *N* HCl for 30 min. The sample was neutralized and a 10 μL aliquot was analyzed for N^2,3-εGua as described below.

A relatively large amount of N^2,3-εGua was produced from the reaction of MCA with deoxyguanosine (Figure 3.20). The amount of adduct formed could not be quantitated since the large amount present saturated the detector (Figure 3.20 top chromatogram). The formation of this adduct from this reaction was not reported in the paper by Kronberg *et al* (1992). The difference between these results is most likely due to the much greater sensitivity of the GC/HRMS method.

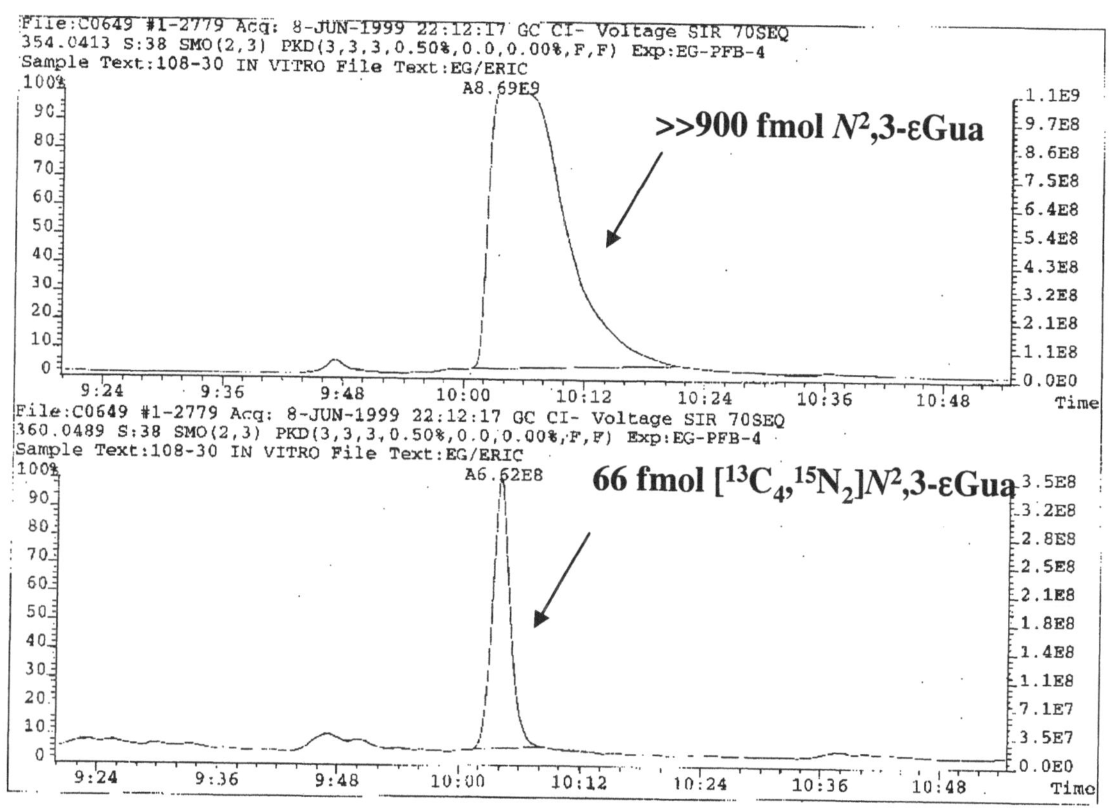

Figure 3.20. Formation of N^2,3-εGua (top) from the reaction of 2'-dGuo with MCA. The large amount of N^2,3-εGua present in this sample saturated the detector. The internal standard is also shown (bottom).

In a follow-up experiment, we administered MCA to 12 F344 rats (100mg/kg). MCA was dissolved in DMSO, mixed with corn oil, and administered by gavage. Animals were sacrificed at 4, 12, or 24 hours following exposure. We analyzed hepatic DNA for increases in N^2,3-εGua, an adduct that is normally present in DNA from lipid peroxidation, using the newly developed GC/HRMS method (Detailed Materials and Methods).

N^2,3-εGua was measured at similar concentrations in control and treated rats. As shown in Figure 3.21, only a small amount of N^2,3-εGua, corresponding to the endogenous presence, was measured in male F344 rats treated with a single corn oil gavage treatment of MCA. However, adduct concentrations were not increased relative to controls (Table 3.10). Although the *in vitro* results of Kronberg *et al* (1992) and our experiments suggest that this adduct is formed *in vitro,* the DNA samples from rats treated with MCA did not show any evidence of N^2,3-εGua.

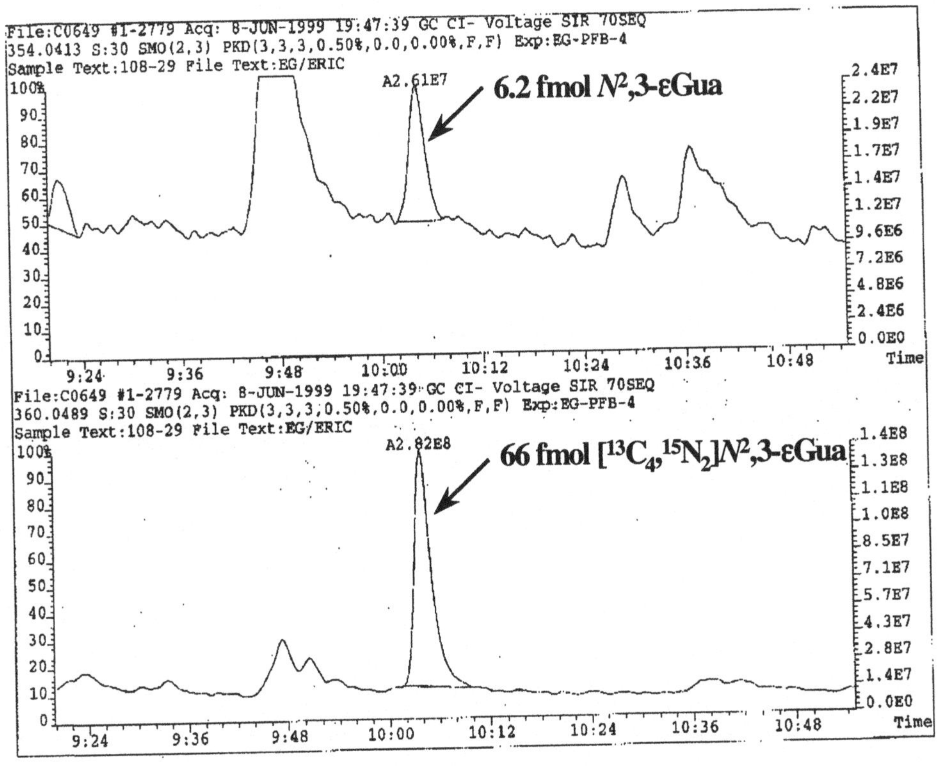

Figure 3.21. Quantitate N^2,3-εGua in rat liver DNA after MCA exposure. N^2,3-εGua was not induced in rat DNA. The small amount of N^2,3-εGua (top) is due to the endogenous presence of this adduct in rat DNA. The internal standard is also shown (bottom).

Table 3.10
The effect of MCA on N^2,3-εGua. Male F344 rats were administered MCA (0 or 100 mg/kg, 0.6 mmol/kg) by corn oil gavage and sacrificed after 4, 12, or 24 hours.

Exposure	N^2,3-ethenoguanine (mol/10^8 mol guanine)
Control	4.6 ± 2.8
4 h	3.6 ± 1.3
12 h	3.0 ± 1.2
24 h	3.4 ± 1.2

Mean ± SD with $p > 0.05$

Quantitate AP Sites In Rat Liver DNA

AP sites are the most common form of endogenous DNA damage. These sites are formed by chemical depurination of normal and damaged DNA bases, by glycosylase activity against DNA adducts and normal DNA bases, and by direct oxidative attack of DNA. AP sites are mutagenic and cleaved AP sites are also clastogenic. Our laboratory has undertaken significant efforts to develop highly sensitive assays for the quantitation and characterization of AP sites (Nakamura *et al.*, 1998 and 2000).

We analyzed hepatic DNA from rats exposed to MCA. F344 rats were administered MCA (100 mg/kg) by corn oil gavage and sacrificed 4, 12 or 24 hours later. Additional rats were exposed to MCA (100mg/kg) daily for 5 days, and sacrificed 4 hours after the last exposure. There was a statistically significant increase in AP site formation only at the 24 hour time point. Subsequently, an acute dose response study was conducted where male F344 rats (n=6 per group) were exposed to a single dose of MCA at 0, 10, 30, 100 or 300 mg/kg dissolved in distilled deionized water via gavage. The major focus of this experiment was to measure the amount of endogenous AP sites in genomic DNA to better understand the formation and repair of AP sites at basal levels after exposure to MCA. Unexpectedly 24h after treatment, all the rats in the high dose group and one of the 100 mg/kg animals were found dead. In addition there was no statistically significant difference, in the amount of AP sites, between the control and MCA exposed animals. All dose groups were evaluated (control, 10, 30 and 100 mg/kg MCA exposures). The mean ± SD values obtained are depicted in Figure 3.22. However, the gavage vehicle between this study and the previous study was different. In the first study MCA was dissolved in DMSO and mixed with corn oil while in the second study MCA was dissolved in water alone.

We are currently planning a larger study, which is outside the scope of the current AwwaRF research project, to determine if the type of gavage vehicle plays a role when animals are exposed to MCA. In addition, we plan to conduct an extensive necropsy where we will collect tissue samples for histology, perform serum chemistry and harvest frozen sections for the measurement of DNA adducts and other lesions to better understand why these animals died.

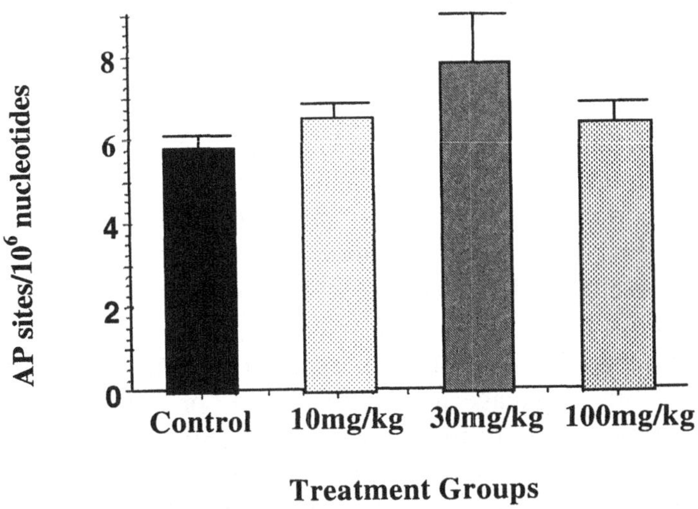

Figure 3.22. Quantitation of AP sites in liver DNA from F344 rats exposed to 0, 10, 30 or 100 mg/kg MCA via a single gavage. Animals were sacrificed 24 hours later. Data represents mean ± SD.

Quantitate 8-Oxoguanine By HPLC-ECD In Rat Liver DNA

A simpler method that promises to limit artificial oxidation was developed by Beckman *et al* (2000). This assay is short with increased sensitivity and allows for direct analysis of oxidized guanine following selective enzymatic digestion by the *Escherichia coli* base excision repair enzyme (Fpg). After Fpg digestion 8-oxoguanine is released from full-length DNA, is separated from high molecular weight molecules by ultrafiltration (10 kDa exclusion) and then analyzed by HPLC-ECD (Detailed Materials and Methods).

F344 liver DNA samples (acute dose response study) from all dose groups (control, 10, 30 and 100 mg/kg MCA exposures) were each mixed with the Fpg enzyme, incubated for 1 hour and subjected to filtration. The filtrate was loaded into HPLC injection vials and stored at -80°C until analyzed by HPLC-ECD for 8-oxoguanine. The retentate from filtration was treated with acid and the resulting hydrolysate was loaded into HPLC injection vials and stored at -80°C until analyzed by HPLC-UV for guanine quantitation.

No detectable amounts of 8-oxoguanine were observed in the control and all the dose groups analyzed (Table 3.11). The level of 8-oxoguanine in these animals was below our detection limit of 5 8-oxoG/10^7G when 25 μg of DNA was applied onto the column.

Table 3.11
Quantitation of 8-oxoguanine by HPLC-ECD in rat liver DNA. F344 rats received a single gavage dose of 0, 10, 30 or 100 mg/kg MCA and were sacrificed 24 hours later.

Treatment	Control	10 mg/kg MCA	30 mg/kg MCA	100 mg/kg MCA
8oxoG/10^6 G	ND	ND	ND	ND

ND = non-detection
Detection Limit: 5 8-oxoG/10^7 G with 25 μg DNA applied onto the column.

Conclusions

MCA is one of the genotoxic halogenated hydroxyfuranones, which are by-products produced by the action of chlorine on humic waters and drinking water. Concerns have also been expressed over an additional risk to human health from pulp and paper mill effluent discharges, containing by-products, into drinking water sources when chlorine is used as bleaching agent (La Londe *et al.*, 1992). In addition, MCA is an intermediate product in the production of the herbicide Pyramin®, which is used against broadleaf weeds and some annual grasses in sugar beet and red table beet crops. Considerable safety measures have been taken to minimize direct exposure to MCA during manufacturing processes since secondary burns can develop in the area of skin contamination as well as other systemic changes (Kolesar and Romancik, 1975).

Although a large amount N^2,3-εGua was formed from the reaction of MCA with deoxyguanosine *in vitro*, the same adduct was not induced following *in vivo* exposure. Therefore, although N^2,3-εGua can be readily formed by direct chemical reaction, this adduct may not be relevant to MCA-induced carcinogenesis. Our results highlight the importance of performing *in vivo* studies on suspect carcinogens such as MCA in drinking water in conjunction with *in vitro* studies.

Considerable progress is also being made with the development of a new immuno slot-blot method for detecting oxidized purines in DNA by using the Oggl enzyme in combination with the AP site assay. 8-OHdG is repaired by Oggl which is a bifunctional DNA glycosylase that cleaves the N-glycosylic bond in DNA leaving an AP site. We plan to repeat quantitation of AP sites and Oggl sites from liver DNA of rats exposed in the study described under Future Research Directions.

Future Research Directions

We are currently planning a larger study, which is outside the scope of the current AwwaRF research project, to determine if the type of gavage vehicle plays a role in toxicity when animals are exposed to MCA. Marked differences in toxicity were evident between the two studies conducted during this research, with much greater toxicity when MCA was administered in water compared with DMSO/corn oil. We plan to conduct single gavage exposures with sacrifice after 24 hours and single daily gavage exposures for 5 days comparing these vehicles. In addition, we plan to conduct an extensive necropsy where we will collect tissue samples for histology, serum chemistry and frozen sections for the measurement of DNA adducts and other lesions to better understand why the rats died in the acute dose response study. Our objective is to focus initially on the liver and kidney and to include an extensive clinical chemistry panel.

The halogenated hydroxyfuranones, like MCA, have not been studied in detail, but may be a potential source of public health risk. To better understand possible mechanisms of MCA-induced genotoxicity, we used a highly sensitive slot-blot assay to characterize aldehydic DNA lesions (ADLs), which include AP sites, and to quantitate 8-oxoguanine (8oxoG; marker for direct base oxidation) by HPLC-ECD *in vivo* after MCA exposure. Five dose groups, each with 6 male F344 rats, received a single gavage dose of 0, 10, 30, 100 or 300 mg/kg MCA. Initial results indicated that there were no significant differences in ADLs and 8oxoG between control and exposed rat liver DNA. We are repeating these analyses and using a new assay where oxidized purines are excised from the DNA backbone by Ogg1, resulting in AP sites detectable by our slot-blot method (Developed by Jun Nakamura). In additon, we used a newly developed, sensitive assay which indirectly monitors the status of DNA single strand break (SSB) repair enzymes by circuitously measuring poly(ADP-ribose) polymerase (PARP) activation through depletion of intracellular NAD(P)H. PARP detects, for example, direct SSBs arising from sugar damage or spontaneous cleavage at AP sites. We applied this assay, which was developed by Jun Nakamura, to XRCC1-proficient and XRCC1-deficient isogenic Chinese hamster ovary cells exposed to MCA. A water-soluble tetrazolium salt was used to monitor NAD(P)H concentrations in living cells through its reduction to a yellow dye detected spectrophotometrically. After 4h, XRCC1-deficient cells contained significantly less NAD(P)H than proficient cells. To distinguish whether the reduction of NAD(P)H was due to decreased mitochondrial function or due to NAD+ depletion by PARP activation, we co-exposed XRCC1-deficient cells to MCA and a PARP inhibitor. The MCA-induced decrease in intracellular NAD(P)H was almost completely blocked, which strongly suggests that the decrease was primarily due to PARP activation through formation of SSBs.

REVISED RESEARCH PLAN – SPECIFIC AIM 11.

Identify and Quantitate DNA Adducts and Lesions in Eker Rat Kidney DNA After Exposure to BDCM—E. Bodes, J. Nakamura, K. S. McDorman, B. F. Pachkowski and J. A. Swenberg

Overview

An additional study was conducted at the US EPA (Dr. Wolf, US EPA/NHEERL, Research Triangle Park, NC) where a total of 56 two month old male Long-Evans (LE) rats were used for this study; 4 male LE rats carrying a *Tsc2* insertion mutation (Eker mutation) and 4 male LE noncarrier sibling rats in each dose group. All confirmed *Tsc2* mutant and wild-type rats used in this study were from MD Anderson Health Sciences Center in Smithville, TX. Animals were dosed for three weeks (21 days) via drinking water with individual water disinfection by-products at high doses previously established to cause cancer with chronic exposure. Dose groups consisted of control, MX (MX, 0.07 g/L), BDCM (BDCM, 0.7 g/L) and potassium bromate ($KBrO_3$; 0.4 g/L) exposure. After three weeks of exposure the rats were weighed, euthanized by carbon dioxide asphyxiation, exsanguinated through cardiac blood collection and necropsied. All tissues were examined for gross abnormalities. Samples were collected, placed in individual containers, frozen in liquid nitrogen and stored at -80°C.

The animal study has been completed and all tissues were collected for morphological and molecular studies. Analysis of water samples during the study indicated that the target concentrations of the chemicals were reached. Chemical consumption data indicated that the tissue dose was similar to the previous Eker rat study.

The goal of this specific aim was to measure 8-oxoguanine and AP sites as biomarkers of oxidative DNA damage in kidneys of rats chronically exposed to disinfection by-products. We used a HPLC-ECD method to identify and quantitate 8-oxoguanine and a sensitive and specific probe slot-blot assay to measure AP site formation. In addition to what was proposed in the "Revised Research Plan" we also conducted quantitative real time reverse transcription-polymerase chain reaction (QRT RT-PCR) experiments to compare mRNA expression of relevant base excision repair genes between groups of rats exposed to different water disinfection by-products.

Quantitate 8-Oxoguanine Levels In Kidney DNA

We have analyzed kidney tissue samples from control, potassium bromate, MX and BDCM exposed Eker rats. Only treatment with potassium bromate caused a treatment related increase in 8-oxoguanine (Table 3.12). Potassium bromate is a known oxidant and has been shown repeatedly to induce the formation of 8-OHdG in rat kidneys (Umemura *et al.*, 1998).

Table 3.12
Quantitation of 8-oxoguanine in kidney DNA from control, potassium bromate, MX and BDCM exposed Eker rats. Eker rats were exposed for 21 days.

Treatment Group	Control	0.4 g/L $KBrO_3$	0.07 g/L MX	0.7 g/L BDCM
8oxoG mol/10^6 G	ND	3.2 ± 0.5	ND	ND

ND = non-detection

Detection Limit: 5 8-oxoG/10^7G using 50 μg of DNA

Quantitate AP Sites In Kidney DNA

DNA from eight (n=8) kidney samples from control and treated animals were evaluated. No increases in abasic sites were observed with treatment (data not shown), but a decrease was apparent in the rats treated with BDCM, suggesting enhanced base excision repair (Figure 3.23).

Conclusions

$KBrO_3$is a by-product from ozonation of high-bromide surface water for production of drinking water and is a rodent carcinogen that produces thyroid, mesothelial and renal tumors. The proposed mechanism of $KBrO_3$ renal carcinogenesis involves the formation of 8-OHdG, a promutagenic base lesion in DNA typically removed through base excision repair. Treatment with $KBrO_3$ produced measurable increases of 8-oxoguanine in the kidney, and no other single chemical treatment caused measurable increases of 8-oxoguanine.

No increases in abasic sites were observed with the different individual treatments, but there was a significant decrease in the rats treated with BDCM which may suggest enhanced base excision repair. However, the expression levels of base excision repair genes analyzed by QRT RT-PCR (data not shown) were not significantly higher than control. Even though the promutagenic DNA lesion 8-oxoguanine was demonstrated to accumulate in tissue after KBrO3 exposure, only, the predicted increase in abasic sites and induction of base excision repair enzyme expression was not observed. These data support the idea that base excision repair of 8-oxoguanine is most dependent on initial glycosylase recognition and cleavage, and that the downstream base excision repair system following this initial step is more than adequate to handle the increased damage.

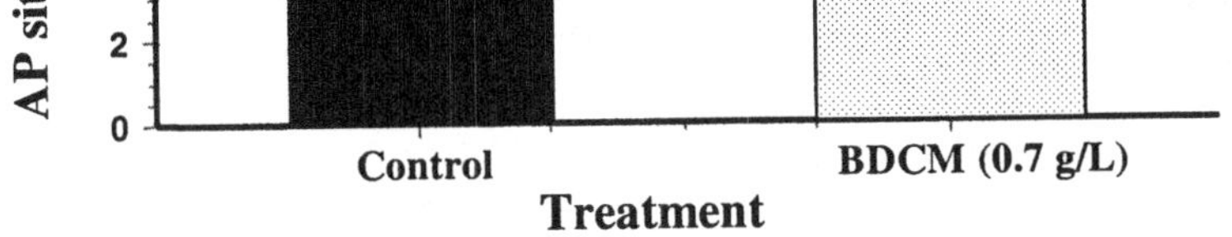

Figure 3.23. Quantitate AP sites in kidney DNA from control and BDCM exposed Eker rats. Mean ± SD with $^*p < 0.05$.

CHAPTER 4: FINAL CONCLUSIONS

E. Bodes and J. A. Swenberg

Chlorine disinfection of drinking water results in the formation of mutagenic and carcinogenic by-products. Although extensive investigations of these chemicals have been conducted, mechanisms underlying their carcinogenic effects remain poorly understood. The major objectives of this research were to: (1) Characterize and quantitate DNA adducts derived from disinfection by-products (DBPs), including BDCM and chlorinated hydroxyfuranones. (2) Study the effects of exposure to DBPs on endogenous DNA adducts. (3) Examine the formation and type of abasic sites in DNA that result from exposure to DBPs. The research should improve our understanding of the mechanisms of DBP toxicity and improve human risk assessment.

Direct DNA adduct formation was examined by exposing BDCM, MCA or MX to nucleosides or DNA. While N^2,3-ethenoguanine readily formed *in vitro* with MCA, none was detectable in MCA-treated rat liver. The yield of DNA adducts formed *in vitro* with glutathione and BDCM was low. Likewise, MX did not readily form ethenoformyl guanine adducts *in vitro* and no adducts were detected in rats chronically exposed to MX.

We hypothesized that increased numbers of endogenous DNA adducts would be present in DBP-treated animals due to depletion of normal cellular defenses. Endogenous DNA adducts were examined in tissues from rodents exposed to MX, MCA, BDCM and DCA and controls. No increases in endogenous adducts were found. Abasic sites are non-coding lesions in DNA that arise following oxidative stress and during DNA repair. We examined tissues from rodents treated with BDCM, MX, DCA and MCA. No increases in AP sites were found in rodents treated with BDCM, MX, or DCA. MCA increased AP sites in liver compared with controls. The AP sites were primarily cleaved 5' to the AP site, suggesting they are not due to reactive oxygen species. Rather, they may be the result of interference with the removal of the 5'-dRp moiety by DNA polymerase β. How MCA interferes with this function of polymerase β is not known, but it may be the rate limiting step for shot-patch base excision repair.

Several disinfection by-products have been shown to be carcinogenic in experimental animals. Likewise, epidemiology studies have suggested increased risks of cancer associated with drinking of chlorinated drinking water. The long-term objective of this research is to better understand the health risks associated with DBPs. We demonstrated that several carcinogenic DBPs do not form direct DNA adducts and do not enhance endogenous adducts. Thus, they are not similar to well known genotoxic carcinogens. This may have important implications for risk assessment.

CHAPTER 5: RECOMMENDATIONS TO UTILITIES

E. Bodes and J. A. Swenberg

Several disinfection by-products have been shown to be carcinogenic in experimental animals. Likewise, epidemiology studies have suggested increased risks of cancer associated with drinking chlorinated water. The long-term objective of this research was to better understand the health risks associated with DBPs. We demonstrated that several carcinogenic DBPs do not form direct DNA adducts and do not enhance endogenous adducts that arise from oxidative stress and lipid peroxidation. Thus, they are not similar to many well known genotoxic carcinogens. This may have important implications for risk assessment. Risk assessments on conventional genotoxic carcinogens are usually based on linear extrapolation in order to be public health protective. This research applied methods similar to those used for many of the classical genotoxic carcinogens. We were not able to identify any direct DNA adducts that arise from covalent binding of DBPs to DNA. Likewise, oxidative stress, a second mechanism thought to be important for this class of chemicals was investigated. This mechanism was postulated to be important for understanding high dose animal cancer studies, since such treatment might overwhelm the body's defense mechanisms. We did not find any evidence for oxidative DNA damage or damage arising from lipid peroxidation following administration of high doses of DBPs.

While it would have been much more satisfying to find positive results in this research, we believe important information has been gathered. The research does not support DBPs as working through classical mechanisms of genotoxicity to induce mutations and cancer. It also does not totally refute such a mechanism. A weight of the evidence approach to analyzing the data suggests that many of the responses observed under the extremely high doses used in animal carcinogenicity studies may bear little relevance to human exposure. Clarification will only come with additional research. This group of chemicals is proving more difficult to understand than many classical carcinogens. Recent data from related studies on Eker rats that examined issues of additivity of low doses of DBPs for risk assessment do not support current regulatory positions. No evidence for additivity was found at the low doses evaluated. These doses were still much greater than those found in chlorinated water. For higher doses, additivity also could not be rejected. Such research is expensive and requires a strong experimental design. While additional research is needed to further clarify the mechanisms involved in DBP carcinogenesis and additivity, it is clear that current regulatory policy is public health conservative and that the actual risk for humans may be less.

REFERENCES

Alhonen-Raatesalmi, A. and Hemminki, K. 1991. Assay for Nucleoside and Nucleotide Binding of a Potent Mutagen, 3-Chloro-4-(Dichloromethyl)-5-Hydroxy-2(5H)-Furanone. *Toxicol. Lett.*, 56:167-172.

Anders, M.W., Dekant, W., and Vamvakas, S. 1992. Glutathione-Dependent Toxicity. *Xenobiotica*, 22:1135-1145.

Anderson, M.E., Clewell, H.J., III, Gargas, M.L., MacNaughton, M.G., McKenna, M.J., Nolan, R.J., and Reitz, R.H. 1991. Physiologically Based Pharmacokinetic Modeling With Dichloromethane, its Metabolite, Carbon Monoxide, and Blood Carboxyhemoglobin in Rats and Humans. *Toxicol. Appl. Pharmacol.*, 108:14-27.

Austin, E.W., Okita, J.R., Okita, R.T., Larson, J.L., and Bull, R.J. 1995. Modification of Lipoperoxidative Effects of Dichloroacetate and Trichloroacetate is Associated With Peroxisome Proliferation. *Toxicology*, 97:59-69.

Austin, E.W., Parrish, J.M., Kinder, D.H., and Bull, R.J. 1996. Lipid Peroxidation and Formation of 8-Hydroxydeoxyguanosine From Acute Doses of Halogenated Acetic Acids. *Fundam. Appl. Toxicol.*, 31:77-82.

Beckman, K.B., Saljoughi, S., Mashiyama, S.T., and Ames, B.N. 2000. A Simpler, More Robust Method for the Analysis of 8-Oxoguanine in DNA. *Free Radic. Biol. Med.*, 29(3/4):357-367.

Beranek, D.T., Weis, C.C., and Swenson, D.H. 1980. A Comprehensive Quantitative Analysis of Methylated and Ethylated DNA Using High Pressure Liquid Chromatography. *Carcinogenesis*, 1:595-606.

Blount, B.C. and Duncan, M.W. 1997. Trace Quantification of the Oxidative Damage Products, Meta- and Ortho-Tyrosine, in Biological Samples by gas Chromatography-Electron Capture Negative Ionization Mass Spectrometry. *Anal. Biochem.*, 244:270-276.

Breen, A.P. and Murphy, J.A. 1995. Reactions of Oxyl Radicals With DNA. *Free Rad. Biol. Med.*, 18:1033-1077.

Bronstein, S.M., Skopek, T.R., and Swenberg, J.A. 1992. Efficient Repair of O^6-Ethylguanine, but not O^4-Ethylthymine or O^2-Ethylthymine, is Dependent Upon O^6-Alkylguanine-DNA Alkyltransferase and Nucleotide Excision Repair Activities in Human Cells. *Cancer Research*, 52:2008-2011.

Brunborg, G., Holme, J.A., Soderlund, E.J., Hongslo, J.K., Vartiainen, T., Lotjonen, S., and Becher, G. 1991. Genotoxic Effects of the Drinking Water Mutagen 3-Chloro-4-(Dichloromethyl)-5-Hydroxy-2[*5H*]-Furanone (MX) in Mammalian Cells in Vitro and in Rats in Vivo. *Mutat. Res.*, 260:55-64.

Bull, R.J., Birnbaum, L.S., Cantor, K.P., Rose, J.B., Butterworth, B.E., Pegram, R., and Tuomisto, J. 1995. Water Chlorination: Essential Process or Cancer Hazard? *Fundam. Appl. Toxicol.*, 28:155-166.

Carro-Ciampi, G., Hunt, P.G., Turner, C.J., and Wells, P.G. 1988. A High-Performance Liquid Chromatographic Assay for Reduced and Oxidized Glutathione in Embryonic, Neonatal, and Adult Tissues Using a Porous Graphite Electrochemical Detector. *J. Pharmacol. Meth.*, 19:75-83.

Chang, L.W., Daniel, F.B., and DeAngelo, A.B. 1991. DNA Strand Breaks Induced in Cultured Human and Rodent Cells by Chlorohydroxyfuranones--Mutagens Isolated From Drinking Water. *Teratog. Carcinog. Mutag.*, 11:103-114.

Chaudhary, A.K., Nokubo, M., Reddy, G.R., Yeola, S.N., Morrow, J.D., Blair, I.A., and Marnett, L.J. 1994. Detection of Endogenous Malondialdehyde-Deoxyguanosine Adducts in Human Liver. *Science*, 265:1580-1582.

Cheng, K.C., Cahill, D.S., Kasai, H., Nishimura, S., and Loeb, L.A. 1992. 8-Hydroxyguanine, an Abundant Form of Oxidative DNA Damage, Causes G→T and A→C Substitutions. *J. Biol. Chem.*, 267:166-172.

Chung, F.L., Young, R., and Hecht, S.S. 1984. Formation of Cyclic 1,N^2-Propanodeoxyguanosine Adducts in DNA Upon Reaction With Acrolein or Crotonaldehyde. *Cancer Res.*, 44: 990-995.

Coffin JC, Ge RR, Yang SM, Kramer PM, Tao LH and Pereira MA. 2000. Effect of trihalomethanes on cell proliferation and DNA methylation in female B6C3F1 mouse liver. Toxicological Sciences 58(2):243-52, Dec, 2000.

Crowley, J.R., Yarasheski, K., Leeuwenburg, C., Turk, J., and Heinecke, J.W. 1998. Isotope Dilution Mass Spectrometric Quantification of 3-Nitrotyrosine in Proteins and Tissues is Facilitated by Reduction to 3-Aminotyrosine. *Anal. Biochem.*, 259:127-135.

Daniel, F.B., DeAngelo, A.B., Strober, J.A., Olson, G.R., and Page, N.P. 1992. Hepatocarcinogenicity of Chloral Hydrate, 2-Chloroacetaldehyde, and Dichloroacetic Acid in the Male B6C3F1 Mouse. *Fundam. Appl. Toxicol.*, 19:159-168.

DeAngelo, A.B., Daniel, F.B., McMillan, L., Wernsing, P., and Savage, R.E., Jr. 1989. Species and Strain Sensitivity to the Induction of Peroxisome Proliferation for Chloroacetic Acids. *Toxicol. Appl. Pharmacol.*, 101:285-298.

DeAngelo, A.B., Daniel, F.B., Most, B.M., and Olson, G.R. 1996. The Carcinogenicity of Dichloroacetic Acid in the Male Fischer-344 rat. *Toxicology*, 114: 207-221.

Dechert, S. and Dekant, W. 1996. The Possible Role of S-chloromethylglutathione in the Genotoxicity of Dichloromethane. *Toxicologist*, 30:93.

Dedon, P.C., Plastaras, J.P., Rouzer, C.A., and Marnett, L.J. 1998. Indirect Mutagenesis by Oxidative DNA Damage: Formation of the Pyrimidopurinone Adduct of Deoxyguanosine by Base Propenal. *Proc. Natl. Acad. Sci.*, 95:11113-11116.

Degan, P., Shigenaga, M.K., Park, E.M., Alperin, P.E., and Ames, B.N. 1991. Immunoaffinity Isolation of Urinary 8-Hydroxy-2'-Deoxyguanosine and 8-Hydroxyguanine and Quantitation of 8-Hydroxy-2'-deoxyguanosine in DNA by Polyclonal Antibodies. *Carcinogenesis*, 12:865-871.

Dekant, W., and Vamvakas, S. 1993. Glutathione-Dependent Bioactivation of Xenobiotics. *Xenobiotica*, 23:873-887.

DeMarini, D.M., Abu-Shakra, A., Felton, C.F., Patterson, K.S., and Shelton, M.L. 1995. Mutation Spectra in Salmonella of Chlorinated, Chloraminated, or Ozonated Drinking Water Extracts: Comparison to MX. *Environ. Mol. Mutagen.*, 26:270-285.

Demple, B., Harrison, L., Wilson, D.M., Bennett, R.A.O., Takagi, T. and Scione, A.G. 1997. Regulation of Eukaryotic Abasic Endonucleases and Their Role in Genetic Stability. *Environ. Health Perspect.*, 105:931-934.

Dragan, Y.P., Hully, J.R., Nakamura, J., Mass, M.J., Swenberg, J.A., and Pitot, H.C. 1994. Biochemical Events During Initiation of rat Hepatocarcinogenesis. *Carcinogenesis*, 15:1451-1458.

Dunnick, J.K., Eustis, S.L., and Lilja, H.S. 1987. Bromodichloromethane, a Trihalomethane That Produces Neoplasms in Rodents. *Cancer Res.*, 47:5189-5193.

El-Ghissassi, F., Barbin, F., Nair, A., and Bartsch, H. 1995. Formation of 1,N^6-Ethenoadenine and 3,N^4-Ethenocytosine by Lipid Peroxidation Products and Nucleic Acid Bases. *Chem. Res. Toxicol.*, 8:278-283.

Fawell, J., Robinson, D., Bull, R., Birnbaum, L.L., Boorman, G., Butterworth, B., Daniel, P., Galal-Gorchev, H., Hauchman, F., Julkunen, P., Klaassen, C., Krasner, S., Orme-Zavaleta, J., Reif, J., and Tardiff, R. 1997. Disinfection By-products in Drinking Water: Critical Issues in Health Effects Research. *Environ. Health Perspect.*, 105:108-109.

Fedtke, N., Boucheron, J.A., Turner, M.J., Jr., and Swenberg, J.A. 1990. Vinyl Chloride-Induced DNA Adducts. I: Quantitative Determination of N^2,3-Ethenoguanine Based on Electrophore Labeling. *Carcinogenesis*, 11:1279-1285.

Floyd RA. 1990. The role of 8-hydroxyguanine in carcinogenesis. Carcinogenesis 11(9):1447-1450.

Franzen, R. and Kronberg, L. 1995. Synthesis of Chlorinated 5-Hydroxy-4-Methyl-2(5H)-Furanones and Mucochloric Acid. *Tetrahedron Lett.*, 36:3905-3908.

Franzen, R., Tanabe, K., and Morita, M. 1998. Isolation of a MX-Guanosine Adduct Formed at Physiological Conditions. *Chemosphere*, 36:2803-2808.

Gao, P., Thornton-Manning, J.R., and Pegram, R.A. 1996. Protective Effects of Glutathione on Bromodichloromethane in Vivo Toxicity and in Vitro Macromolecular Binding in Fischer 344 Rats. *J. Toxicol. Environ. Health*, 49:145-159.

Graves, R.J., Trueman, P., Jones, S., and Green, T. 1996. DNA Sequence Analysis of Methylene Chloride-Induced HPRT Mutations in Chinese Hamster Ovary Cells: Comparison With the Mutation Spectrum Obtained for 1,2-Dibromoethane and Formaldehyde. *Mutagenesis*, 11:329-333.

Ham, A-J., Ranasinghe A., Morinello, E.J., Nakamura, J., Upton, P.B., Johnson, F., and Swenberg, J.A. 1999. Immunoaffinity/gas Chromatography/High-Resolution Mass Spectrometry Method for the Detection of N^2,3-Ethenoguanine. *Chem. Res. Toxicol.*, 12:1240-6.

Ham. A.J., Ranasinghe, A., Koc, H., and Swenberg, J.A. 2000. 4-Hydroxy-2-nonenal and Ethyl Linoleate Form N^2,3-Ethenodeoxyguanosine Under Peroxidizing Conditions. *Chem. Res. Toxicol.*, 13:1243-1250.

Hartley, D.P., Kroll, D.J., and Petersen, D.R. 1997. Prooxidant-Initiated Lipid Peroxidation in Isolated rat Hepatocytes: Detection of 4-Hydroxynonenal- and Malondialdehyde-Protein Adducts. *Chem. Res. Toxicol.*, 10:895-905.

Helbock, H.J., Beckman, K.B., Shigenaga, M.K., Walter, P.B., Woodall, A.A., Yeo, H.C., and Ames, B.N. 1998. DNA Oxidation Matters: the HPLC-Electrochemical Detection Assay of 8-Oxo-deoxyguanosine and 8-Oxo-guanine. *Proc. Natl. Acad. Sci.*, 95:288-293.

Hildesheim, M.E., Cantor, K.P., Lynch, C.F., Dosemeci, M., Lubin, J., Alvanja, M., and Craun, G. 1998. Drinking Water Source and Chlorination By-products. II. Risk of Colon And Rectal Cancers. *Epidemiology*, 9:29-35.

Hofer, T. and Moller, L. 1998. Reduction of Oxidation During the Preparation of DNA and Analysis of 8-Hydroxy-2'-deoxyguanosine. *Chem. Res. Toxicol.*, 11:882-887.

Holme, J.A., Haddeland, U., Haung, K., and Brunborg, G. 1999. DNA Damage Induced by the Drinking Water Mutagen 3-Chloro-4-(Dichloromethyl)-5-Hydroxy-2(5H)-Furanone (MX) in Mammalian Cells in Vitro and in Mice. *Mutat. Res.*, 441:145-153.

Huymphreys, W.G., Kim, D.H., Cmarik, J.L., Shimada, T., and Guengerich, F.P. 1990. Comparison of DNA-Alkylating Properties and Mutagenic Responses of a Series of S-(2-Haloethyl)-Substituted Cysteine and Glutathione Derivatives. *Biochemistry*, 29:10342-10350.

Hyttinen, J.M. and Jansson, K. 1995. PM2 DNA Damage Induced by 3-Chloro-4-(Dichloromethyl)-5-Hydroxy-2(5H)-Furanone (MX). *Mutat. Res.*, 348:183-186.

Hyttinen, J.M.T. and Jansson, K. 1996. Kinds of Mutations Induced by 3-Chloro-4-(Dichloromethyl)-5-Hydroxy-2-(5H)-Furanone (MX) in the Hprt Gene of Chinese Hamster Ovary Cells. *Carcinogenesis*, 17:1179-1181.

Hyttinen, J.M., Nittykoski, M., Jansson, K. 1995. Lack of Uniformity in the Mutational Spectra of Chlorohydroxyfuranones in Salmonella typhimurium Strain TA 100. *Mutagenesis*, 10:321-323.

International Agency for Research on Cancer (IARC). 1991. IARC Monographs on the Evaluation of Carcinogenic Risks to Humans. Vol. 52. Chlorinated Drinking Water; Chlorinated By-Products; Some Other Halogenated Compounds; Cobalt and Cobalt Compounds. IARC, Lyon, France.

International Agency for Research on Cancer (IARC) 1995. IARC Monographs on the Evaluation of Carcinogenic Risks to Humans. Vol. 63. dry Cleaning, Some Chlorinated solvents and Other Industrial Chemicals. IARC, Lyon, France.

Jorgenson, T.A., Meierhenry, E.F., Rushbrook, C.J., Bull, R.J., and Robinson, M. 1985. Carcinogenicity of Chloroform in Drinking Water to Male Osborne-Mendel Rats and Female B6C3F1 Mice. *Fundam. Appl. Toxicol.*, 5:760-769.

Kasai H. 1997. Analysis of a form of oxidative DNA damage, 8-hydroxy-2'-deoxyguanosine, as a marker of cellular oxidative stress during carcinogenesis. Mutat Res 387:147-163.

Keegan, T.E., Simmons, J.E., and Pegram R.A. 1998. NOAEL and LOAEL Determinations of Acute Hepatotoxicity for Chloroform and Bromodichloromethane Delivered in an Aqueous Vehicle to F344 Rats. *J. Toxicol. Environ. Health.*, 55:65-75.

Kluwe, W.M. and Hook, J.B. 1981. Potentiation of Acute Chloroform Nephrotoxicity by the Glutathione Depletor Diethyl Maleate and Protection by the Microsomal Enzyme Inhibitor Piperonyl Butoxide. *Toxicol. Appl. Pharmacol.*, 59:457-466.

Knasmuller, S., Zohrer, E., Kronberg, L., Kundi, M., Franzen, R., and Schulte-Hermann, R. 1996. Mutational Spectra of Salmonella typhimurium Revertants Induced by Chlorohydroxy-furanones, By-products of Chlorine Disinfection in Drinking Water. *Chem. Res. Toxicol.*, 9:374-381.

Kolesar D and Romancik V. 1975. Clinical and laboratory findings in the acute damage by precursors of the weed-killer "burex". Bratisl Lek Listy. Aug;64(2):132-40.

Komulainen, H., Kosma, V.M., Vaittinen, S.L., Vartiainen, T., Kaliste-Korhonen, E., Lotjonen, S., Tuominen, R.K., and Tuomisto, J. 1997. Carcinogenicity of the Drinking Water Mutagen 3-Chloro-4-(Dichloromethyl)-5-Hydroxy-2(5H)-Furanone in the rat. *J. Natl. Cancer Inst.*, 89:848-856.

Kronberg, L., Munter, T., Maki, J., Le Curieux, F., Pluskota, D., and Sjoholm, R. 1999. Adducts of Chlorohydroxyfuranones and of Aldehyde Conjugates. *IARC Sci. Publ.* 150:115-121.

Kronberg, L., Karlsson, S., and Sjoholm, R. 1993. Formation of Ethenocarbaldhyde Derivatives of Adenosine and Cytidine in Reactions With Mucochloric Acid. *Chem. Res. Toxicol.*, 6:495-499.

Kronberg, L., Sjoholm, R., and Karlsson, S. 1992. Formation of 3,N^4-Ethenocytidine, 1,N^6-Ethenoadenosine and 1,N^2-Ethenoguanosine in Reactions of MCA With Nucleosides. *Chem. Res. Toxicol.*, 5:852-855.

Kuchino, Y., Mori, F., Kasai, H., Inoue, H., Iwai, S., Miura, K., Ohtsuka, E., and Nishimura, S. 1987. Misreading of DNA Templates Containing 8-Hydroxydeoxyguanosine at the Modified Base and at Adjacent Residues. *Nature*, 327:77-79.

La, D.K., Lilly, P.D., Anderegg, R.J., and Swenberg, J.A. 1995. DNA Adduct Formation in B6C3F1 Mice and Fischer-344 Rats Exposed to 1,2,3-Trichloropropane. *Carcinogenesis*, 16:1419-1424.

La, D.K. and Swenberg, J.A. 1996. DNA adducts: biological markers of exposure and potential applications to risk assessment. *Mutat Res.* 1996 Sep; 365(1-3):129-46.

La, D.K., Schoonhoven, R., Ito, N., and Swenberg, J.A. 1996. The Effects of Exposure Route on DNA Adduct Formation and Cellular Proliferation by 1,2,3-Trichloropropane. *Toxicol. Appl. Pharmacol.*, 140:108-114.

La, D.K. and Swenberg, J.A. 1997. The Induction of Endogenous DNA Adducts in Tissues of Fischer-344 Rats Following Gavage Administration of 1,2,3-Trichloropropane. *Proc. Am. Assoc. Cancer Res.*, 38:131.

LaLonde RT, Leo H, Perakyla H, Dence CW, Farrell RP. 1992. Associations of the bacterial mutagenicity of halogenated 2(5H)-furanones with their MNDO-PM3 computed properties and mode of reactivity with sodium borohydride. *Chem Res Toxicol.* May-Jun;5(3):392-400

Larson, J.L. and Bull, R.J. 1992. Metabolism and Lipoperoxidative Activity of Trichloroacetate and Dichloroacetate in Rats and Mice. *Toxicol. Appl. Pharmacol.*, 115:268-277.

Larson, J.L., Wolf, D.C., and Butterworth, B.E. 1994. Induced Cytotoxicity and Cell Proliferation in the Hepatocarcinogenicity of Chloroform in Female B6C3F1 Mice: Comparison of Administration by Gavage in Corn oil vs. ad Libitum in Drinking Water. *Fundam. Appl. Toxicol.*, 22:90-102.

Le Curieux, F., Munter, T., and Kronberg, L. 1997. Identification of Adenine Adducts Formed in Reaction of Calf Thymus DNA With Mutagenic Chlorohydroxyfuranones Found in Drinking Water. *Chem. Res. Toxicol.* 10(10):1180-1185.

Leeuwenburgh, C., Hansen, P., Shaish, A., Holloszy, J.O., and Heinecke, J.W. 1998. Markers of Protein Oxidation by Hydroxyl Radical and Reactive Nitrogen Species in Tissues of Aging Rats. *Am. J. Physiol.*, 274:R453-R461.

Leuratti, C., Singh, R., Lagneau, C., Farmer, P.B., Plastaras, J.P., Marnett, L.J., and Shuker, D.E. 1998. Determination of Malondialdehyde-Induced DNA Damage in Human Tissues Using an Immunoslot-blot Assay. *Carcinogenesis*, 19:1919-1924.

Lilly, P.D., Ross, T.M., and Pegram, R.A. 1997. Trihalomethane Comparative Toxicity: Acute Renal and Hepatic Toxicity of Chloroform and Bromodichloromethane Following Aqueous Gavage. *Fundam. Appl. Toxicol.*, 40:101-10.

Marsteinstredet, U., Brunborg, G., Bjoras, M., Soderlund, E., Seeberg, E., Kronberg, L., and Holmes, J.A. 1997. DNA Damage Induced by 3-Chloro-4-(Dichloromethyl)-5-Hydroxy-2[5H]-Furanone (MX) in HL-60 Cells and Purified DNA in Vitro. *Mutat. Res.*, 390:171-178.

Meier, J.R., DeAngelo, A.B., Daniel, F.B., Schenck, K.M., Skelly, M.F., and Jaung, S.L. 1989. DNA Adduct Formation in Bacterial and Mammalian Cells by 3-Chloro-4-(Dichloromethyl)-5-Hydroxy-2(5H)-Furanone (MX). *Environ. Mol. Mutagen.*, 14(Suppl. 15):128.

Melnick RL, Kohn MC, Dunnick JK and Lwininger JR. 1998. Regenerative hyperplasia is not required for liver tumor induction in female B6C3F1 mice exposed to trihalomethanes. Toxicol Appl Pharmacol 148(1)137-47.

Mills, C.J., Bull, R.J., Cantor, K.P., Reif, J., Hrudey, S.E., and Huston, P. 1998. Health Risks of Drinking Water Chlorination By-Products: Report of an Expert Working Group. *Chronic Diseases in Canada*, 19:91-102.

Mol, C.D., Hosfield, D.J., and Tianer, J.A. 2000. Abasic Site Recognition by two Apurinic/Apyrimidinic Endonuclease Families in DNA Base Excision Repair: the 3' Ends Justify the Means. *Mutat. Res.*, 460:211-229.

Morinello, E.J., Ham, A.J., Ranasinghe, A., Sangaiah, R., and Swenberg, J.A. 2001. Simultaneous Quantitation of N^2,3-Ethenoguanine and 1,N^2-Ethenoguanine With an Immunoaffinity/gas Chromatography/High Resolution Mass Spectrometry Assay. *Chem. Res. Toxicol.*, 14:327-334.

Moriya, M., Ou, C., Bodepudi, V., Johnson, F., Takeshita, M., and Grollman, A.P. 1991. Site-Specific Mutagenesis Using a Gapped Duplex Vector: a Study of Translation Synthesis Past 8-Oxodeoxyguanosine in E. coli. *Mutat. Res.*, 254:218-288.

Morris, R.D., Audet, A.M., Angelillo, I.F., Chalmers, T.C., and Mosteller, F. 1992. Chlorination, Chlorination By-products, and Cancer: a Meta-analysis. *Am. J. Public Health.* 82:955-963.

Munter, T., Le Curieux, F., Sjoholm, R., and Kronberg, L. 1999. Identification of an Ethenoformyl Adduct Formed in the Reaction of the Potent Bacterial Mutagen 3-Chloro-4-(Dichloromethyl)-5-Hydroxy-2(*5H*)-Furanone With Guanosine. *Chem. Res. Toxicol.*, 12:46-52.

Munter, T., Le Curieux, F., Sjohlm, R., and Kronberg, L. 1998. Reaction of the Potent Bacterial Mutagen 3-Chloro-4--(Dichloromethyl)-5-Hydroxy-2(*5H*)-Furanone (MX) With 2'-Deoxyadenosine and Calf Thymus DNA: Identification of Fluorescent Propenoformyl Derivatives. *Chem. Res. Toxicol.*, 11:226-233.

Misra, R.R., Chiang, S.Y., and Swenberg, J.A. 1994. A Comparison of two Ultrasensitive Methods for Measuring 1,N^6-Etheno-2'-deoxyadenosine and 3,N^4-Etheno-2'-deoxycytidine in Cellular DNA. *Carcinogenesis*, 15:1647-1652.

Nakamura, J., La, D.K., and Swenberg, J.A. 2000. 5'-Nicked Apurinic/Apyrimidinic Sites are Resistant to β-Elimination by β-Polymerase and are Persistent in Human Cultured Cells After Oxidative Stress. *J. Biol. Chem.*, 275:5323-5328.

Nakamura, J. and Swenberg, J.A. 1999. Endogenous Apurinic/Apyrimidinic Sites in Genomic DNA of Mammalian Tissues. *Cancer Res.*, 59:2522-2526.

Nakamura, J., Walker, V.E., Upton, P.B., Chiang, S.Y., Kow, Y.W., and Swenberg, J.A. 1998. Highly Sensitive Apurinic/Apyrimidinic Site Assay can Detect Spontaneous and Chemically Induced Depurination Under Physiological Conditions. *Cancer Res.*, 58:222-225.

Nakae, D., Mizumoto, Y., Kobayashi, E., Noguchi, O., and Konishi, Y. 1995. Improved Genomic/Nuclear DNA Extraction for 8-Hydroxydeoxyguanosine Analysis of Small Amounts of rat Liver Tissue. *Cancer Lett.*, 97:233-239.

Nath, R.G., Ocando, J.E., Richie, J.P., Jr., and Chung, R.L. 1997. Effect of Glutathione Depletion on Exocyclic Adduct Levels in the Liver DNA of F344 Rats. *Chem. Res. Toxicol.,* 10:1250-1253.

National Cancer Institute (NCI). 1976. Carcinogenesis Bioassay of Chloroform. NCI, Bethesda, MD.

National Toxicology Program (NTP). 1987. Toxicology and Carcinogenesis Studies of Bromodichloromethane (CAS No. 75-27-4) in F344/N Rats and B6C3F1 Mice. *Toxicology*, 110:103-111.

Nishikawa, A., Kinae, N., Furukawa, F., Mitsui, M., Enami, T., Hasegawa, T., and Takahashi, M. 1994. Enhancing Effect of 3-Chloro-4-(Dichloromethyl)-5-Hydroxy-2[*5H*]-Furanone (MX) on Cell Proliferation and Lipid Peroxidation in the rat Gastric Mucosa. *Cancer Lett.*, 85:151-157.

Nunn, J.W. and Chipman, J.K. 1994. Induction of DNA Strand Breaks by 3-Chloro-4-(Dichloromethyl)-5-Hydroxy-2[*5H*]-Furanone and Humic Substances in Relation to Glutathione and Calcium Status in Human White Blood Cells. *Mutat. Res.*, 341:133-140.

Parrish, J.M., Austin, E.W., Stevens, D.K., Kinder, D.H., and Bull, R.J. 1996. Haloacetate-Induced Oxidative Damage to DNA in the Liver of Male B6C3F1 Mice. *Toxicology*, 110:103-111.

Pegram, R.A., Anderson, M.E., Warren, S.H., Ross, T.M., and Claxton, L.D. 1997. Glutathione *S*-Transferase-Mediated Mutagenicity of Trihalomethanes in Salmonella typhimurium: Contrasting Results With Bromodichloromethane and Chloroform. *Toxicol. Appl. Pharmacol.*, 144:183-188.

Pogozelski, W.K. and Tullius, T.D. 1998. Oxidative Strand Scission Of Nucleic Acids: Routes Initiated by Hydrogen Abstraction From the Sugar Moiety. *Chem. Rev.*, 98:1089-1107.

Ramana, C.V., Boldogh, I., Izumi, T., and Mitra, S. 1998. Activation Of Apurinic/Apyrimidinic Endonuclease in Human Cells by Reactive Oxygen Species and its Correlation With Their Adaptive Response to Genotoxicity of Free Radicals. *Proc. Natl. Acad. Sci. U. S. A.* 95:5061-5066.

Requena, J.R., Fu, M.X., Ahmed, M.U., Jenkins, A.J., Lyons, T.J., and Thorpe, S.R. 1996. Lipooxidation Products as Biomarkers of Oxidative Damage to Proteins During Lipid Peroxidation Reactions. *Nephrology, Dialysis, Transplantation*, 11(Suppl. 5):48-53.

Requena, J.R., Fu, M.S., Ahmed, M.U., Jenkins, A.J., Lyons, T.J., Baynes, J.W., and Thorpe, S.R. 1997. Quantification of Malondialdehyde and 4-Hydroxynonenal Adducts To Lysine Residues in Native and Oxidized Human Low-Density Lipoprotein. *Biochem. J.*, 322:317-325.

Richter, C., Park, J.W., and Ames, B.N. 1998. Normal Oxidative Damage to Mitochondrial and Nuclear DNA is Extensive. *Proc. Natl. Acad. Sci.*, 85:6465-6467.

Roussyn, I., Yen, T.Y., Wond, V.A., Cattley, R.C., Cunningham, M.L., and Swenberg, J.A. 1998. Wy-14,643, A Potent Peroxisome Proliferator, Induces Expression of N-Methylpurine-DNA Glycosylase (MPG) in rat Liver and Increases Urinary Excretion of 1,N^6-Ethenoadenine. *Advances in DNA Damage and Repair-Oxygen Radical Effects, Cellular Protection, and Biological Consequences*. Dizadaroglu, M. and Karakaya A. (eds.), Plenum Press, New York.

Schroder, K.R., Hallier, E., Peter, H., and Bolt, H.M. 1992. Dissociation of a new Glutathione S-Transferase Activity in Human Erythrocytes. *Biochem. Pharmacol.*, 43:1671-1674.

Schut, H.A.J., Cummings, D.A., Lin, E.L.C., and Daniel, F.B. 1991. ^{32}P-Postlabeling Analysis of DNA Adduct Formation of the Drinking Water Mutagen 3-Chloro-4-(Dichloromethyl)-5-Hydroxy-2(*5H*)-Furanone (MX). *Environ. Mol. Mutagen.*, 17(Suppl. 19):65-66.

Sobol, R.W., Prasad, R., Evenski, A., Baker, A., Yang, X.P., Horton, J.K., and Wilson, S.H. 2000. The Lyase Activity of the DNA Repair Protein Beta-Polymerase Protects From DNA-Damage-Induced Cytotoxicity. *Nature*, 405:807-810.

Stevens, J.L. and Anders, M.W. 1981. Metabolism of Haloforms to Carbon Monoxide. IV. Studies on the Reaction Mechanism in Vivo. *Chem. Biol. Interact.*, 37:365-374.

Talpaert-Borle M. 1989. Formation, detection and repair of AP sites. Mutat Res 181:45-56.

Testai E, Di Marzio S and Vittozzi L. 1990. Multiple activation of chloroform in hepatic microsomes from uninduced B6C3F1 mice. Toxicol Appl Pharmacol 104:496-503.

Their, R., Taylor, J.B., Pemble, S.E., Humphreys, W.G., Persmark, M., Ketterer, B., and Guengerich, F.P. 1993. Expression of Mammalian Glutathione S-Transferase 5-5 in Salmonella typhimurium TA 1535 Leads to Base-Pair Mutations Upon Exposure to Dihalomethanes. *Proc. Natl. Acad. Sci.*, 90:8576-8580.

Umemura, T., Takagi, A., Sai, K., Hasegawa, R., and Kurokawa, Y. 1998. Oxidative DNA Damage and Cell Proliferation in Kidneys of Male and Female Rats During 13-Weeks Exposure to Potassium Bromate ($KBrO_3$). *Arch. Toxicol.*, 72(5):264-269.

Vaittinen, S.L., Komulainen, H., Kosma, V.N., Julkunen, A., Maki-Paakkanen, J., Jansson, K., Vartiainen, T., and Tuomisto, J. 1995. Subchronic Toxicity of 3-Chloro-4-(Dichloromethyl)-5-Hydroxy-2(5H)-Furanone (MX) in Wistar Rats. *Food Chem. Toxicol.*, 33:1027-1037.

Wood, M.L., Dizdaroglu, M., Gajewski, E., and Essigmann, J.M. 1990. Mechanistic Studies of Ionizing Radiation and Oxidative Mutagenesis: Genetic Effects of a Single 8-Hydroxyguanine (7-Hydro-8-oxoguanine) Residue Inserted at a Unique Site in a Viral Genome. *Biochemistry*, 29:7024-7032.

Yi, D., Smythe, G.A., Blount, B.C., and Duncan, M.W. 1997. Peroxynitrite-Mediated Nitration of Peptides: Characterization of the Products by Electrospray and Combined gas Chromatography-Mass Spectrometry. *Arch. Biochem. Biophys.*, 344:253-259.

ABBREVIATIONS

amu	anatomic mass units
ANOVA	Analysis of variance
ARP	aldehyde reactive probe
AP	apurinic/apyrimidinic
APE	apurinic/apyrimidinic endonucleases
APE1	AP endonuclease 1
ASB	abasic site slot-blot assay
Awwa	American Water Works Association
AwwaRF	Awwa Research Foundation
BDCM	bromodichloromethane
BHT	butylated hydroxytoluene
BSA	bovine serum albumin
°C	degrees Celsius
C18	octadecyl silica
CID	collision-induced dissociation
Corp	Corporation
DBPs	disinfection by-products
DCA	dichloroacetate or dichloroacetic acid
dA	deoxyadenosine
dC	deoxycytidine
dG	deoxyguanosine
dGuo	deoxyguanosine
DMSO	dimethyl sulfoxide
DNA	deoxyribose nucleic acid
DNase	deoxyribonuclease
ddH_2O	deionized distilled water
ddwater	deionized distilled water
dRp	deoxyribose phosphate
5'-dRp	5'-deoxyribophosphodiester
E. coli	*Escherichia coli*
EfdGuo	3-(2'-deoxy-β-D-ribofuranosyl)-7-formylimidazo[1,2-α]purin-9(4H)-one
ESI	electrospray ionization
eV	electron volts
Exo III	*E. coli* exonuclease III (3'-5' exonuclease)
εdA	ethenodeoxyadenosine
εdC	ethenodeoxycytidine
εG	ethenoguanine
εfdGuo	ethenoformyldeoxyguanosine
εGua	ethenoguanine

ESI/MS	electrospray ionization with mass spectrometry
F344	Fischer-344
fmol	femtomole
Fpg	formamidopyrimidine glycosylase
g	gram
g/L	gram per liter
GC	gas chromatography
GC to TA	guanine-cytosine to thymine-adenine transversion
GC/HRMS	gas chromatography/ high-resolution mass spectrometry
GSTT1	glutathione *S*-transferase θ
h	hour
HCl	hydrochloric acid
HEPES/KOH	4-(2-hydroxy ethyl)piperazine-1-ethane sulfonic acid/potassium hydroxide
^{1}H-NMR	proton nuclear magnetic resonance
HPLC	high-pressure liquid chromatography
HPLC-ECD	HPLC with electrochemical detection
HRMS	high-resolution mass spectrometry
IA	immunoaffinity
IA/GC/HRMS	immunoaffinity/gas chromatography/high-resolution mass spectrometry
ILSI	International Life Sciences Institute
Inc.	Incorporated
ISBN	International Standard Book Number
kV	kilovolt
L	liter
LC-MS/MS	liquid chromatography with tandem mass spectrometry
LC-ESI/MS	liquid chromatography with electrospray ionization tandem mass spectrometry
LOAEL	
m	meter
MCA	mucochloric acid
mg	milligram
mg/kg	milligram per kilogram
$MgCl_2$	magnesium chloride
MHz	megahertz
min	minute
mL	milliliter
mL/min	milliliter per minute
mm	millimeter
mM	millimolar
mmol	millimole

mmol/kg	millimole per kilogram
MMS	methylmethane sulfonate
mRNA	messenger ribonucleic acid
mV	millivolt
MX	mutagen X; 3-chloro-4-(dichloromethyl)-5-hydroxy-2-(*5H*)furanone
m/z	mass to charge
μA	microampere
μg	microgram
μg/mL	microgram per milliliter
μg/kg	microgram per kilogram
μL	microliter
μL/min	microliter per minute
μm	micrometer
μM	micromolar
μmol/mol	micromole per mole
N	normal
NC	nitrocellulose
ng/L	nanogram per liter
nm	nanometer
nmol	nanomol
NMR	nuclear magnetic resonance
NOAEL	no observed affect level
NTP	National Toxicology Program
8-OHdG	8-hydroxydeoxyguanosine
ODS	octadecyl silica
OGG1	oxidized guanine glycosylase 1
PAC	Project Advisory Committee
PBS	phosphate buffered saline
PFBBr	pentafluorobenxyl bromide
pmol/mL	picomol per milliliter
ppm	parts per million
psi	pounds per square inch
RNA	ribonucleic acid
RNase	ribonuclease
rpm	revolutions per minute
SD	standard deviation
SE	standard error
SRM	selective reaction monitoring
T7 Exo	T7 gene 6 exonuclease (5'-3' exonuclease)
TEMPO	2,2,6,6-tetramethylpiperidinoxy

TE	Tris-EDTA
US EPA	United States Environmental Protection Agency
UV	ultraviolet
V	Volt
x g	times gravity

IWA Publishing's authorised EU representative for General Product Safety Regulations is Diane D'Arras, 15 rue Duret, 75116 Paris, France, e-mail: safety@iwap.co.uk.

Printed and bound by CPI Group (UK) Ltd, Croydon, CR0 4YY
17/09/2026
02218813-0013